Transfer and Stockpiling Operations in the Crude Oil Refining Industry

Transfer and Stockpiling Operations in the Crude Oil Refining Industry provides an overview of crude oil refining processes highlighting stockpiling operations, which unfortunately tend to be underestimated in comparison with processing units operations like FCC and hydrotreating. The book presents a deep analysis of the current context and challenges imposed on players in the downstream industry and how the stockpiling operations can help the refiners achieve flexibility and better competitive positioning. How stockpiling managers can use their knowledge to call attention to stockpiling assets and manage process safety and performance adequately is described. This book:

- Offers a simplified approach to crude oil refining processes.
- Discusses economic information related to the downstream business and the role of stockpiling operations in the refining business.
- Introduces newer trends in the industry, such as synergy between FCC and hydrocracking technologies as a strategy to maximize profitability and modern naphtha alkylation technologies as a safety route to achieve high quality and low sulfur gasoline.
- Presents challenges related to adequate process safety and asset management in stockpiling operations.
- Describes the main interfaces between the stockpiling operations of a crude oil refinery and relevant areas like midstream companies and how to manage this relation.

Offering a contemporary view of current challenges and opportunities in the downstream oil and gas business, this practical book is aimed at readers working in the fields of petroleum and chemical engineering, calling attention to the relevance of stockpiling operations to the global results of the refineries.

Dr. Marcio Wagner da Silva is Process Engineering Manager at a crude oil refinery based in São José dos Campos, Brazil. He earned a bachelor's in chemical engineering from the University of Maringa (UEM), Brazil, and a PhD in chemical engineering from the University of Campinas (UNICAMP), Brazil. He has extensive experience in research, design, and construction in the oil and gas industry, including developing and coordinating projects for operational improvements and debottlenecking to bottom barrel units. Moreover, Dr. Marcio Wagner earned an MBA in project management from the Federal University of Rio de Janeiro (UFRJ), in digital transformation at Pontifical Catholic University of Rio Grande do Sul (PUC/RS), and in production and operations management at State University of São Paulo (USP) and is certified in business from Getulio Vargas Foundation (FGV).

Transfer and Stockpiling Operations in the Crude Oil Refining Industry

Marcio Wagner da Silva

CRC Press
Taylor & Francis Group
Boca Raton London New York

CRC Press is an imprint of the
Taylor & Francis Group, an **informa** business

Designed cover image: Shutterstock

First edition published 2025
by CRC Press
2385 NW Executive Center Drive, Suite 320, Boca Raton FL 33431

and by CRC Press
4 Park Square, Milton Park, Abingdon, Oxon, OX14 4RN

CRC Press is an imprint of Taylor & Francis Group, LLC

ISBN: 978-1-032-75060-6 (hbk)
ISBN: 978-1-032-75061-3 (pbk)
ISBN: 978-1-003-47221-6 (ebk)

DOI: 10.1201/9781003472216

Typeset in Times
by SPi Technologies India Pvt Ltd (Straive)

Contents

Preface ix
Acknowledgments x

Chapter 1 Current Context of the Downstream Industry 1

1.1 Changing the Focus of the Downstream Business – More Petrochemicals and Less Fossil Fuels 4
1.2 The Growing Pressure for Decarbonizing the Downstream Industry 7
1.3 Biofuel Production in Brazil 9
1.4 Challenges of Renewables Coprocessing in Crude Oil Refineries 11
Bibliography 14

Chapter 2 Overview of Crude Oil Refining Processes 16

2.1 Crude Oil Distillation Units 16
2.1.1 Vacuum Distillation Section 18
2.2 Thermal Conversion Processes 20
2.2.1 Visbreaking Process 20
2.2.2 Delayed Coking Process 23
2.2.3 Solvent Deasphalting Technologies 26
2.3 Catalytic Conversion Processes 30
2.3.1 Fluid Catalytic Cracking Units (FCC) 30
2.4 Catalytic Reforming Technologies 33
2.5 Naphtha Alkylation Technologies 36
2.6 Naphtha Isomerization 41
2.7 Light Paraffin Dehydrogenation 44
2.8 Hydrotreating Processes 46
2.8.1 Naphtha Hydrotreating Technologies 50
2.8.2 Diesel Hydrotreating Units 53
2.8.3 Bottom Barrel Hydrotreating Technologies (Residue Upgrading) 55
2.8.4 Hydrocracking Technologies 60
Bibliography 66

Chapter 3 The Role of Transfer and Stockpiling Operations in a Crude Oil Refinery 68

3.1 Remembering Fluid Mechanics Concepts 69
3.2 Pumps Review 71
3.2.1 The Cavitation Phenomenon 74
3.2.2 The Cavitation Phenomenon 75

3.2.3 Operating Issues of Centrifugal Pumps 77
3.2.4 Operating Issues of Centrifugal Pumps 78
Bibliography 79

Chapter 4 Types and Classification of Storage Tanks 80

4.1 Fixed Roof Tanks 81
4.2 Floating Roof Tanks 81
4.3 Tanks Classification by Bottom Type 84
4.4 Tanks Classification by Shell Type 85
4.5 Typical Accessories of Storage Tanks 85
4.5.1 Venting System 85
4.5.2 Mixing System 86
4.5.3 Gauging System 88
4.5.4 Heating System 89
4.5.5 Draining or Dewatering System 89
4.5.6 Bund 91
4.5.7 Tank Farms 91
4.5.8 Blanketing System 92
4.5.9 Applying the Knowledge 93
Bibliography 93

Chapter 5 Blending Operations (Crude Oil, LPG, Gasoline, Fuel Oil, and Diesel) 94

5.1 Quality Requirements of Crude Oil Derivatives 94
5.2 Blending Operations – Crude Oil 105
5.3 Blending Operations – Gasoline Production 108
5.3.1 Alkylated Naphtha – A Special and Relevant Case 111
5.4 Blending Operations – Diesel Production 116
5.5 Blending Operations – LPG Production 116
5.6 Blending Operations – Fuel Oil Production 117
Bibliography 120

Chapter 6 Flaring Systems and How to Manage these Systems Adequately 121

6.1 Main Components of Flaring Systems 126
6.1.1 Stack Sealing System 126
6.1.2 Ignition System 127
6.1.3 Liquid Separation System 127
6.1.4 Liquid Seal Vessel 128
6.1.5 Flare Header 128
6.1.6 Flare Gas Recovery System 129
6.2 Operating Issues of Flaring Systems 132
6.2.1 Pulsation of Flare Flame 132

6.2.2 Rumbling Phenomena in Elevated Flare Systems.....133
6.3 Chemical Flaring Systems.....133
Bibliography.....139

Chapter 7 Process Safety Management in Transfer and Stockpiling Operations.....140

7.1 Asset Management.....145
7.2 Discipline.....149
7.3 Integration.....154
7.3.1 High- and Low-level Alarm.....157
7.3.2 High- and Low-level Alarm.....157
7.3.3 Dynamic High- or Low-level Alarm.....158
Bibliography.....160

Chapter 8 Calling the Attention to Transfer and Stockpiling Area – The Relation with Maintenance.....161

Bibliography.....168

Chapter 9 Asset Integrity Management in Transfer and Stockpiling Sector.....169

9.1 The Corrosion Threat – A Brief Review.....170
9.1.1 Corrosion Classification and Forms.....171
9.2 Corrosion Protection.....176
9.3 The Threat of "Dead Legs".....178
9.4 Corrosion Process in Storage Tanks.....180
9.5 The Threat of Fluid Hammer in Stockpiling Assets.....187
Bibliography.....187

Chapter 10 Managing the Interface between Refinery Stockpiling Operations and Middle Stream Companies.....188

10.1 Managing the Threat of the Midstream Companies to the Refinery.....188
10.2 The Mutual Operations between Midstream and Crude Oil Refineries.....189
10.3 The Interface between the Refining Production Planning and Scheduling and the Stockpiling Area.....191
Bibliography.....195

Chapter 11 Process Safety and Asset Management in LPG and Liquefied Gas Storage Farm.....196

11.1 Liquefied Petroleum Gases (LPGs).....196
11.2 Propylene.....197

11.3 The Hazards to Storage Liquefied Petroleum Gases 199
11.4 Maintenance of LPG Spheres 202
11.5 Contamination of LPG Spheres 204
11.6 Propylene Contamination with Carbonyl Sulfide (COS) 212
11.7 Water Flooding System of Liquefied Gases Storage Farm 215
Bibliography 217

Chapter 12 Management of Metering and Billing Operations 218

12.1 Ensuring the Accuracy and Reliability of the Metering Operations 220
Bibliography 222

Chapter 13 Special Topics in Crude Oil Refining 223

13.1 Synergy between the FCC and Hydrocracking Processing Units as a Strategy to Maximize Refining Margins 223
13.1.1 Fluid Catalytic Cracking Technologies – The Carbon Rejection Route 224
13.1.2 Improving the Yield of Petrochemicals in the Refining Hardware – The Petrochemical FCC Technologies 226
13.1.3 Deep Hydrocracking Technologies – The Hydrogen Addition Route 228
13.1.4 Petrochemical Integration – Strong Dependence of High Bottom Barrel Conversion 229
13.1.5 The Crude Oil to Chemicals – What Is the Role of FCC and Hydrocracking? 231
13.1.6 Synergy between Hydrocracking and FCC Units – A High Bottom Barrel Conversion Refining Hardware 231
13.1.7 The Side Effect of Cracked Feeds – A Special Challenge to Hydrocracking Units 233
13.2 Meeting the Gasoline Quality and Process Safety Requirements through New Naphtha Alkylation Technologies 235
13.2.1 Gasoline Production Process – The Relevance of Octane Boosters 237
13.2.2 Traditional Technologic Routes to Naphtha Alkylation 238
13.2.3 Alternatives and Safer Technologies to Naphtha Alkylation 242
Bibliography 244

Index 245

Preface

Why a Book about Stockpiling Operations?

I was stockpiling manager of a great crude oil refinery for exactly three years. I found an adverse scenario from all points of view: asset management, operational discipline, process safety, and, more importantly, team morale… During the 1,095 days that I was stockpiling manager, I asked the same question every day: how was it possible to achieve this scenario? After a long period of reflection, I understand that the problem is that the refinery personnel and managers do not have complete knowledge of the relevance of stockpiling operations to a crude oil refinery. The impacts and the risk scenarios tend to be underestimated in relation to processing units like distillation and catalytic cracking, leading to a poor policy of asset management and other management systems in the downstream business. This is provoked by a kind of management myopia which tends to pay less attention to the stockpiling of assets than the refining processing units, and this culture tends to reach all areas of a crude oil refinery, from the maintenance to operation teams in all downstream companies.

This feeling is reinforced by the fact that there are no books focusing on stockpiling operations and so many about crude oil refining processes (including one from this author). Faced with this, I decided to publish this book, aiming to shed light on the relevance of stockpiling operations to a typical crude oil refinery and help to improve the quality of asset management of these operating assets based on my experience.

The book was developed by mixing technical and management arguments in order to make clear the relevance of transfer and stockpiling operations for the crude oil refining business. It is important to highlight that, in the current business environment, professionals must be as technically capable as they are managerially capable. If they do not act as managers, professionals must at least have the management knowledge that allows for an adequate critical analysis of the orders and guidance received from a strategic point of view.

This book aims to help young professionals know the criticality of stockpiling assets in the refining business and develop experienced professionals with respect to this fundamental area to the results of a crude oil refinery.

"As long as supposed loyalty is rewarded at the expense of results, we will continue to see unprepared people in leadership positions"

Acknowledgments

It's impossible to achieve any good result alone, and this book is no different. In this sense, I would like to start by acknowledging and saying thank God! The support that I received from my family, especially my wife Ana Glaucia and my daughter Manuela, was invaluable; this book simply would not exist without their support.

In my developing journey, I had the opportunity to learn from good professors from the State University of Maringa (UEM) and the State University of Campinas (UNICAMP), so I would like to say thanks to all the good professors from the Chemical Engineering Courses of both universities.

1 Current Context of the Downstream Industry

The downstream industry is facing a transitive period where the players of the industry are suffering increasing pressure to minimize their impact on the environment, and this pressure is translated into stricter regulations. The newest threat to refiners is the reduction of the consumer market; in the last years, it became common, news about countries that intend to reduce or ban the production of vehicles powered by fossil fuels in the middle term, mainly in developed markets such as Europe and Asia.

With the improvement in fuel efficiency, the growing market of electric vehicles tends to decline the participation of transportation fuels in the global crude oil demand. New technologies like additive manufacturing (3D printing) have the potential to produce a great impact on transportation demands, leading to even more impact on transportation fuel demand. Furthermore, the higher availability of lighter crude oils favors the oversupply of lighter derivatives that facilitate the production of petrochemicals against transportation fuels as well as the higher added value of petrochemicals in comparison with fuels. Despite the recent forecasts, the transportation fuel demand is still the main revenue driver to the downstream industry, as presented in Figure 1.1, based on data from the International Energy Agency (IEA).

According to Figure 1.1, the transportation fuels demand represents close to five times the demand by petrochemicals as well as a focus on transportation fuels of the current refining hardware. Despite this data, a strong trend of reduction in transportation fuel demand is observed, followed by a growing market of petrochemicals. Still, according to IEA data, presented in Figure 1.2, relevant growth is expected in the petrochemical's participation in the global oil demand.

With the improvement in fuel efficiency, the growing market of electric vehicles tends to decline the participation of transportation fuels in the global crude oil demand. Figure 1.3 presents the growth of electric vehicles in the last years in the global market.

Another threat is the growing participation of renewable raw materials in the crude oil refineries, in response to the society's requirement for energy transition efforts. In the last months, some important players have announced the conversion of some crude oil refineries into renewable processing plants while other players and technology developers have announced the production of diesel and jet fuel by applying coprocessing of crude oil and renewable raw material like HVGO in some refineries around the world.

Another deep change in the downstream sector that reinforces the necessity of high-conversion refining hardware is the IMO 2020. Restrictive regulations like IMO 2020 raised, even more, the pressure on refiners with low bottom barrel conversion capacity once required higher capacity to add value to residual streams, especially

DOI: 10.1201/9781003472216-1

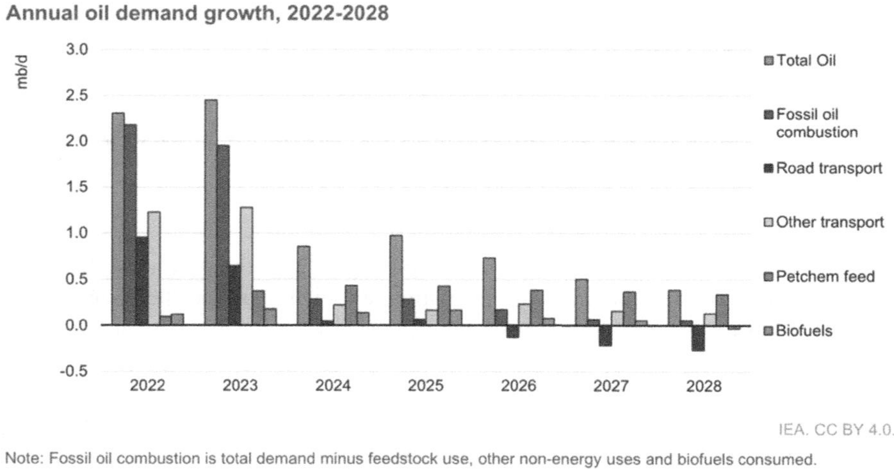

FIGURE 1.1 Forecast for petrochemical feedstock/transportation fuels demand for next years (IEA, 2023). [Note: The top-to-bottom labeling order of "Total Oil, fossil oil combustion…" corresponds to each group of bar graphs going from left to right.]

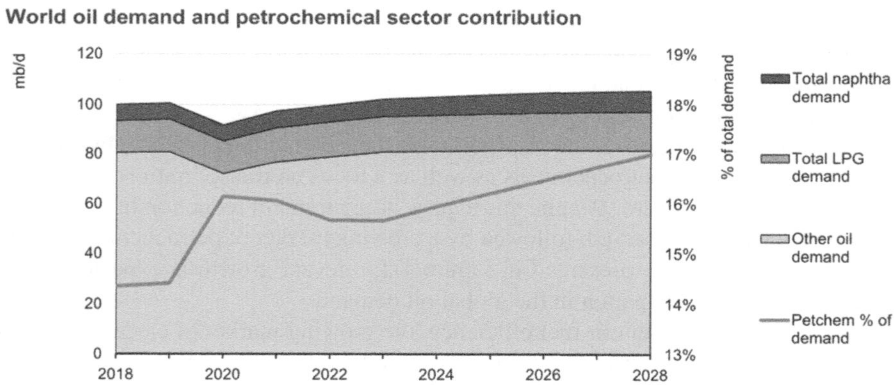

FIGURE 1.2 Change in the profile of global crude oil demand (IEA, 2023). [Note: "Petchem % of demand" corresponds to the line graph running through this figure. The three other color shade labels from top to bottom correspond to the color shade layers of the graph from top to bottom.]

related to sulfur content that was reduced from 3.5% (in mass) to 0.5%. Refiners with easy access to low-sulfur crude oils present a relative competitive advantage in this scenario; these players can rely on relatively low cost residue upgrading technologies to produce the new marine fuel oil (Bunker) as carbon rejection technologies (Solvent Deasphalting, Delayed Coking, etc.), but they are the minority in the market. Most of the players need to look for sources of low sulfur crudes, which present higher costs, putting pressure on their refining margins, or look for deep bottom barrel conversion technologies to ensure more value addition to processed crude oils and avoid losing

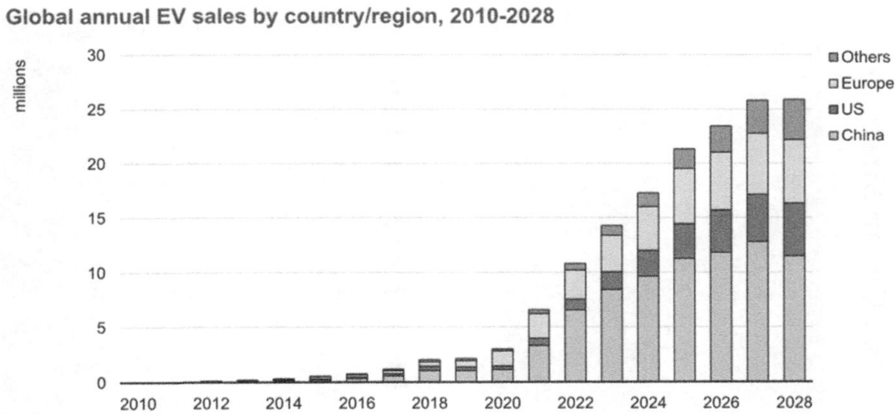

FIGURE 1.3 Growth of the electric vehicles fleet over the years (Global EVs Outlook 2023, IEA) [Note: The color shade labels from top to bottom correspond with each year's bar graph's color shade layers from top to bottom.]

competitiveness in the downstream market. For these refiners, deep residue upgrading, like hydrocracking technologies, can offer great operational flexibility, despite the high capital spending. In this scenario, with the necessity for higher value addition to the bottom barrel stream and the growing market of petrochemicals, refiners with adequate bottom barrel conversion capacity can achieve great competitive advantage in the downstream industry.

Based on the description above, it's possible to apply the article published by W. Chan Kim and Renée Mauborge called "Blue Ocean Strategy" in Harvard Business Review to classify the competitive markets in the downstream industry. In this article the authors define the conventional market as a red ocean where the players tend to compete in the existing market, focusing on defeating competitors through the exploration of existing demand, leading to low differentiation and low profitability. The blue ocean is characterized by looking for space in non-explored (or few explored) markets, creating and developing new demands, and reaching differentiation. This model can be applied (with some specificities, as it is a commodity market) to the downstream industry, considering the traditional transportation fuels refineries and the petrochemical sector.

Due to these characteristics, the transportation fuels market can be imagined like the red ocean, where the margins tend to be low and under high competition between the players with low differentiation capacity. On the other side, the petrochemicals sector can be faced like the blue ocean where few players are able to meet the market in competitive conditions, higher refining margins, and significant differentiation in relation to refiners dedicated to the transportation fuels market. Figure 1.4 presents the basic concept of the blue ocean strategy in comparison with the traditional red ocean, where the players fight for market share with low margins.

As presented above, the market forecasts indicate that the refiners able to maximize petrochemicals against transportation fuels can achieve highlighted economic performance in the short term. In this sense, the crude oil to chemicals technologies

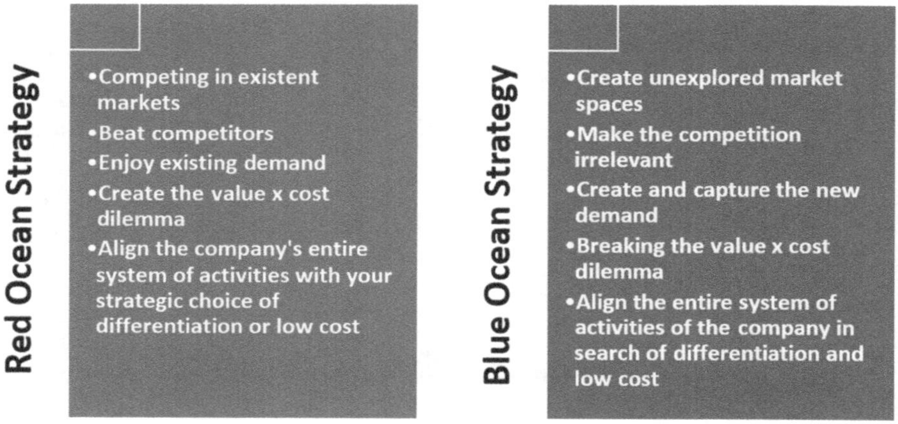

FIGURE 1.4 Differences between blue and red ocean strategies (Based on Kim and Mauborge, 2004).

can offer even more competitive advantage to the refiners with the capacity of capital investment.

It can be difficult for some people to understand the term "differentiation" in the downstream industry, as this is a market that deals with commodities, but the differentiation here is related to the capacity to reach more added value to the processed crude oil and as presented above, nowadays this is translated into the capacity to maximize the petrochemicals yield, creating differentiation between integrated and non-integrated players.

1.1 CHANGING THE FOCUS OF THE DOWNSTREAM BUSINESS – MORE PETROCHEMICALS AND LESS FOSSIL FUELS

In this business environment, it's possible to adapt the Ansoff Matrix to consider the contraction profile of the transportation fuels market to analyze the available alternatives to the downstream players. The Ansoff Matrix is presented in Figure 1.5.

In Figure 1.5, the current position of downstream players is focused on transportation fuel demand that presents a contraction profile as aforementioned. In this scenario, there are three alternatives for the players:

1. Look for new clients: This alternative seems attractive at first glance, but the stricter regulations and trend of reduction in consumption create great pressure over the consumption of fossil fuels. The major consumers of transportation fuels are still in developing economies such as Brazil, Mexico, and India, but the most efficient engines and substitute technologies like hybrid and electric vehicles tend to reduce the market growth even in these countries;
2. Offer a new value addition: In the face of the reduction in transportation fuels, an attractive strategy to the downstream sector is to offer a new proposed

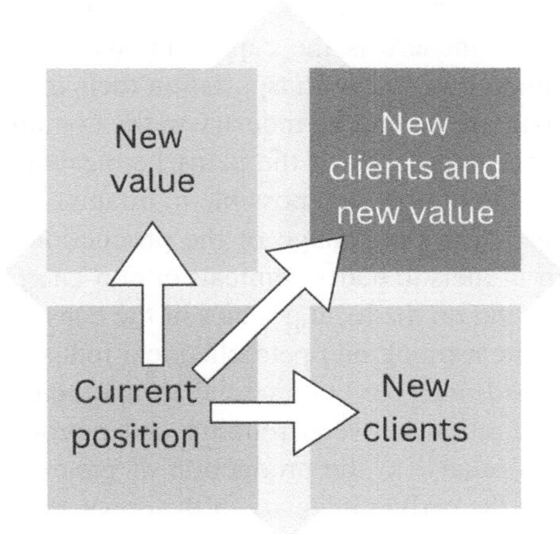

FIGURE 1.5 Adapted Ansoff Matrix to an in-contraction market (Based on Rogers, 2016).

value to the market through higher value addition to the processed crude oils as well as needed materials to the society with a lower environmental footprint than fossil fuels. The petrochemical intermediates have higher added value to refiners and growing demand. As aforementioned data, the substitution of steel in some engineering materials is an interesting way to ensure a market for petrochemicals in the short term; in this sense, the refiners can change the production focus from transportation fuels to petrochemicals, especially in markets like Asia and Europe where the falling demand for transportation fuels is most significant. Beyond the petrochemicals, the capacity to add value to bottom barrel streams appears like a competitive advantage.

3. New clients and new value addition: Strategically, this alternative seems the right way to follow, mainly for refiners with the most complex refining hardware. Through the promotion of closer integration with the petrochemical sector, the refiners not only offer a higher proposed value to the clients and society but can reach a new range of customers capable of ensuring higher added value to the processed crude oils and lower operational costs through available synergies between refining and petrochemical assets.

Another strategic planning model which can be applied to the current scenario of the downstream industry is the PESTEL model. This theory applies six fundamental factors which influence the strategy of any business. These factors are:

- Political: The growing pressure to minimize the environmental footprint of the energetic matrix is leading the governments to raise the pressure on the downstream players to reduce their carbon intensity. This is translated into an increasingly hostile scenario for the fossil fuels.
- Economic: The higher added value of the petrochemicals in comparison with transportation fuels, as well as their higher potential of circularity, are

important drivers to consider in the strategic planning of the downstream players. The question here is the capital intensity required to make the change of a refining asset from transportation fuels to petrochemicals.

- Social: The impact of the energy industry in the communities is high, considering the higher circularity of the petrochemicals against the fuels and the higher economic returns. It's possible to imagine a better social benefit for society considering the change of the production focus in the downstream sector from fuels to petrochemicals or non-energetic derivatives like lubricants depending on the local policies of the consumer market.
- Technological: The refining and petrochemical industries are well technologically developed, and the commercial technologies are totally able to allow closer integration between refining and petrochemical assets. The challenge here is related to the production of green hydrogen on a large scale to minimize the carbon intensity of the downstream sector.
- Environmental: Again, the environmental factor favors the petrochemical sector due to the higher potential of circularity than the transportation fuels.
- Legal: We are seeing growing pressure on the downstream players to reduce the carbon intensity of their operations, which is translated to stricter regulations. This factor overlaps with the political factor, which favors the adoption of petrochemicals against fuels as a production focus.

A simplified analysis based on the PESTEL model can indicate the favorability of changing the production focus from fossil fuels to petrochemicals. This is summarized in Figure 1.6.

At this point, it's important to explain that the energy transition is an incremental and not disruptive process. The change in the production focus from fuels to petrochemicals

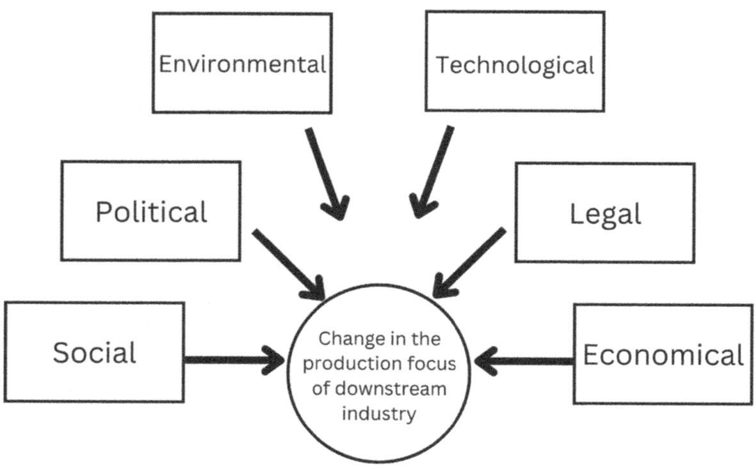

FIGURE 1.6 PESTEL analysis for the change in the production focus of the downstream players.

is not an easy task, and the capital requirement is a barrier among the downstream players. Our point of view is that the players with capital power and inserted in more economically developed markets will change quickly from fuels to petrochemicals, while the players with less access to capital and based in developing economies will keep their focus on transportation fuels.

1.2 THE GROWING PRESSURE FOR DECARBONIZING THE DOWNSTREAM INDUSTRY

The increasing necessity to reduce the environmental impact produced by fossil fuels has created a trend of decarbonization of the energetic matrix at a global level, creating then a new challenge to the crude oil production and processing chain. The current geopolitical crisis due to the war between Ukraine and Russia put another element in this scenario: the necessity to reduce the carbon intensity at the same time to keep and ensure the energy security for the nations.

Under this scenario, one of the available alternatives is raising the renewable fuels participation in the energetic matrix as well as the higher use of renewable raw materials in the feed stream of crude oil refineries, and this fact has led some refining technology licensors to dedicate efforts to develop processes for this purpose.

The adoption of synergies between fossil fuels and renewables in the downstream industry depends on the market where the refiner is inserted, mainly related to the availability of renewable raw materials as well as the capacity of the installed refining hardware to process the renewable streams.

Despite these limitations, it's important to understand that renewables are already a reality in the market, contributing to the reduction of the demand for fossil raw materials. According to data from IEA, the COVID-19 pandemic caused the first contraction in biofuels market in two decades, as presented in Figure 1.7.

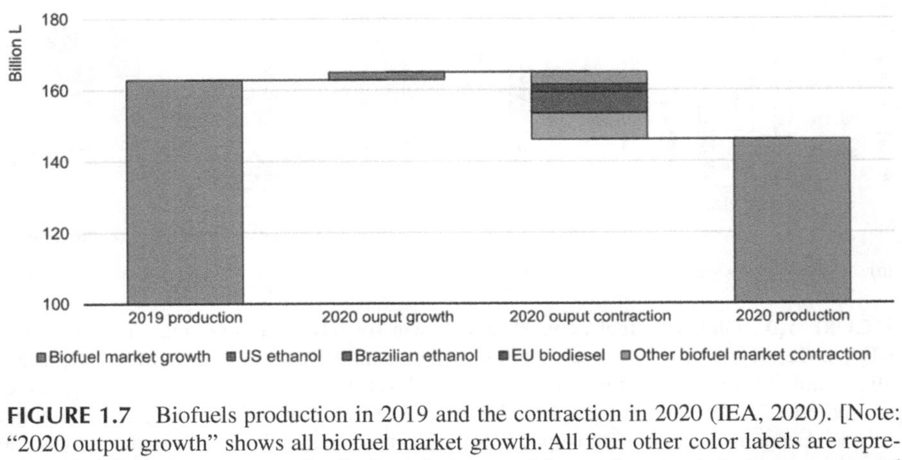

FIGURE 1.7 Biofuels production in 2019 and the contraction in 2020 (IEA, 2020). [Note: "2020 output growth" shows all biofuel market growth. All four other color labels are represented in "2020 output contraction," where the bar graph layers from top to bottom correspond to all other color labels from right to left.]

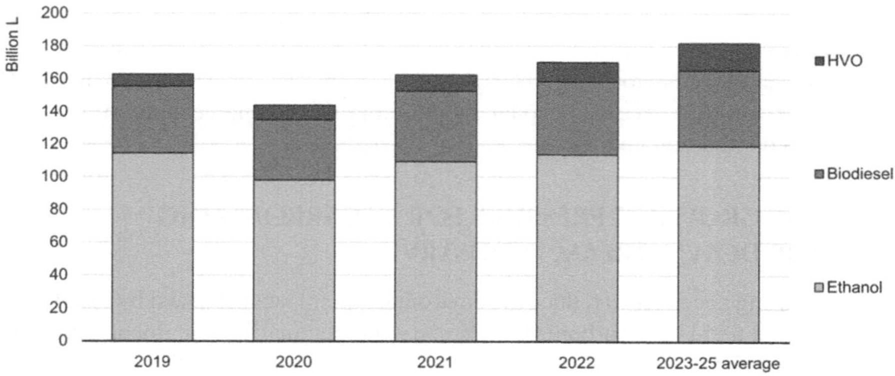

FIGURE 1.8 Global biofuels production forecast (IEA, 2020). [Note: Both the color key labels and each bar graph's layers correspond from top to bottom of the key.]

Despite this contraction in 2020, the stricter regulations and policy pressure tend to drive a fast recovery and expansion in the biofuels demand, still according to data from IEA presented in Figure 1.8.

Considering these trends, it's possible to estimate the impact of biofuels in the crude oil refining industry, and the coprocessing of renewable raw materials in traditional crude oil refineries can be an attractive decarbonization strategy. After the COVID-19 pandemic, some refiners decided to convert some refining assets to process renewable raw materials, reinforcing this trend in the new scenario of the downstream industry. Considering more recent data, the demand for biofuels was confirmed to be consistent growth, as shown in Figure 1.9.

According to the IEA, the United States, Canada, India, Indonesia, and Brazil are responsible for 80% of the biofuel expansion in the period. This fact is strictly related

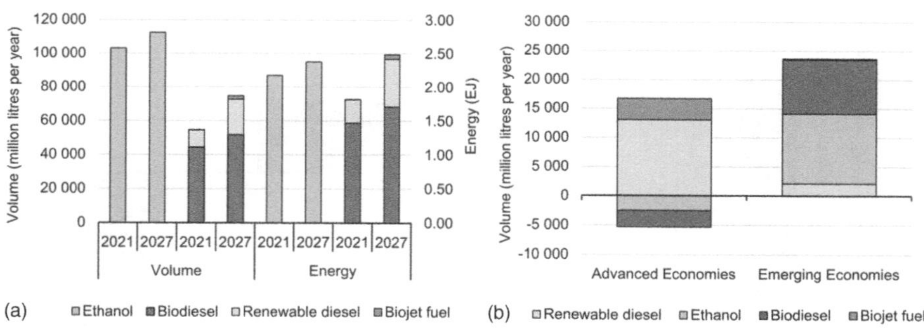

FIGURE 1.9 Global biofuel demand and growth for advanced and emerging economies (IEA, 2022). [Note: (a) Volume and energy graph – both 2021 and 2027's leftmost graphs are all ethanol. 2021's rightmost graphs have a small layer of renewable diesel and a larger layer of biodiesel. 2027's rightmost graphs follow this same trend but with a thin top layer of biojet fuel. (b) Economies graph – from top to bottom advanced economies are biojet fuel, renewable diesel, ethanol, and biodiesel. From top to bottom emerging economies are biodiesel, ethanol, and renewable diesel.]

to established policies, which sustain this growth in these countries. As expected, in developing economies like Brazil and Indonesia, the rising gasoline and diesel consumption boosts biofuel consumption in these countries, while in developed economies such as the United States and Canada, the falling demand for transportation fuels reduces the use of biofuels. As the author has been discussing through the last years, this scenario can lead to a kind of "run" of the big refining players aiming to maximize the added value to the processed crude through closer integration between refining and petrochemical assets in order to maximize petrochemicals against transportation fuels due to the higher added value and growth market of petrochemicals in developed economies; this can also lead to a surplus of gasoline and diesel in developing countries like Brazil and India once the players with low capital power to promote a deep petrochemical integration will look for consumer markets for their surplus production.

1.3 BIOFUEL PRODUCTION IN BRAZIL

Brazil has a long tradition in biofuel production. In 1975, due to the petroleum crisis, the Brazilian authorities launched an alternative fuel program called PROALCOOL, where the main intention was to support the development of ethanol from sugar cane as automobile fuel in substitution of gasoline to reduce the external dependence on the Brazilian energetic matrix.

According to the Brazilian Petroleum Agency (ANP), in 2019, Brazilian ethanol production reached 35.3 million m^3 considering the volumes of anhydrous and hydrated ethanol. This production reveals consistent growth in the production over the years. Figure 1.10 shows the ethanol production profile over the last years in the Brazilian market.

Based on data from Figure 1.10, Brazilian ethanol production grew at an average annual rate of 0.60%. Considering only the anhydrous ethanol, the annual growth is

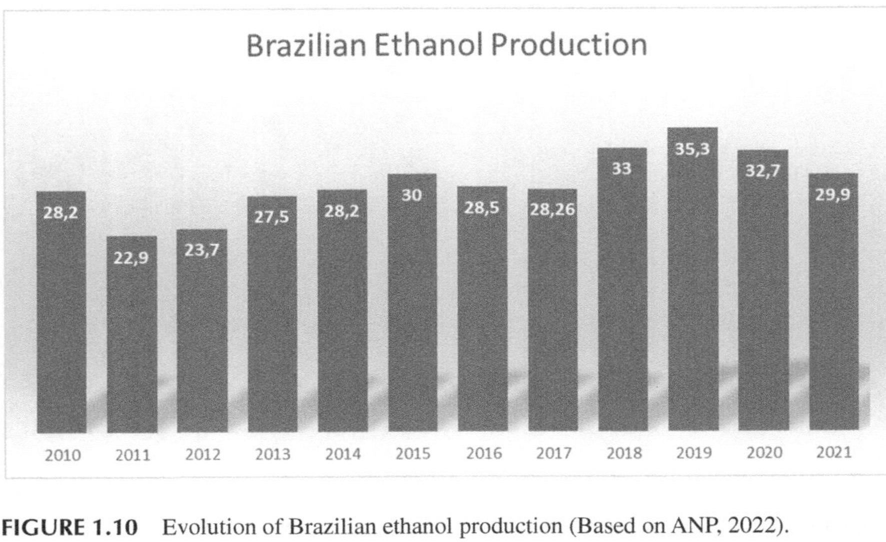

FIGURE 1.10 Evolution of Brazilian ethanol production (Based on ANP, 2022).

even more expressive, reaching 1.68% in the 2012–2021 period. Despite this, in the last two years, there has been a reduction in Brazilian ethanol production, comparing the values in 2020 and 2021. The reduction is around 8.2%. This reduction is due to the valorization of the sugar market, which led some producers to direct their production to sugar to the detriment of ethanol, as well as due to the impact of the COVID-19 pandemic. This fact reveals an important "dilemma" in biofuel production which is frequently discussed in energy forums related to the competition between fuels and food raw materials involving biomass. This is a fundamental question which needs to be adequately managed.

In the same period, the importations of ethanol to the Brazilian market were discouraged due to the devaluation of the Brazilian currency.

By law, the gasoline commercialized in Brazil has 27% in volume of anhydrous ethanol, which is applied to improve the gasoline quality (octane boosting) and to ensure the participation of renewable fuels in the Brazilian energetic matrix. Hydrated ethanol is commercialized in gas stations as pure fuel for automobiles. Still, according to data from ANP, in 2019 the Brazilian production of hydrated ethanol reached 22.4 million m^3 with an average annual growth of 3.8% considering the 2012–2021 period.

Brazil is a great transportation fuel consumer, and the main driver of the Brazilian economy is diesel. Due to the country's dimensions and the transport infrastructure, which relies on road transport, the total production of diesel in the Brazilian market reached close to 43 million m^3 in 2021, a growth of 1.5% in comparison with 2020. By law, diesel fuel commercialized in the Brazilian territory needs to contain 12% volume of biodiesel, and the intention of the Brazilian government is to raise this percentage to 14% in 2024. Figure 1.11 presents the evolution of Brazilian biodiesel production between 2010 and 2021 in million m^3.

The internal production of biodiesel in Brazil shows consistent growth with an average annual growth of 16.9% between 2012 and 2021. The main raw material

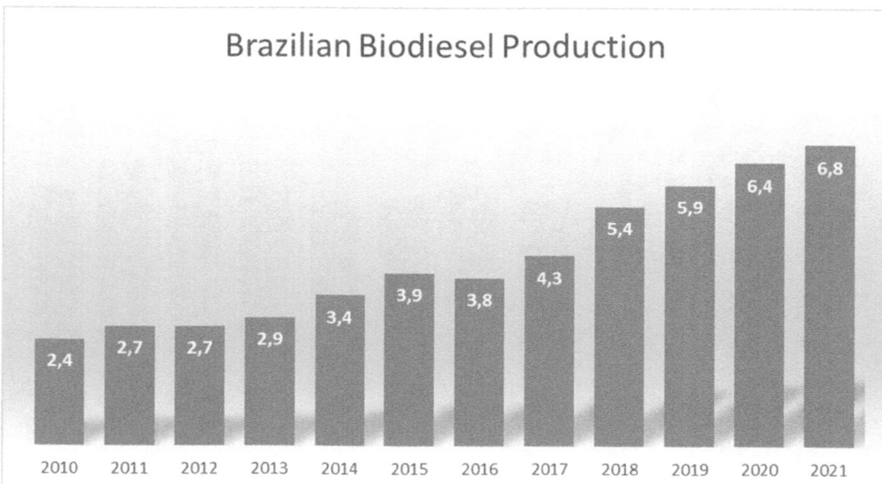

FIGURE 1.11 Evolution of Brazilian biodiesel production (ANP, 2022).

applied to produce biodiesel in Brazil is soybean oil, with close to 72% of the total production, followed by animal fat at 10 %.

As described above, biofuels are fundamental to sustaining the energetic matrix and economic development of Brazil. The blending of anhydrous ethanol into gasoline and biodiesel into diesel represents a kind of strategy to produce cleaner fuels, but this is not the only strategy being applied to the refiners aiming to reduce the environmental footprint of transportation fuels.

An important trend of the energy transition in the downstream industry is the coprocessing of renewable raw materials in the crude oil refineries. This strategy involves feeding the renewable raw material directly to the refining process, which represents a more challenging decarbonization strategy.

1.4 CHALLENGES OF RENEWABLES COPROCESSING IN CRUDE OIL REFINERIES

The use of renewable raw materials in crude oil refineries has been discussed in the last decades. The adoption of synergies between fossil fuels and renewables in the downstream industry depends on the market where the refiner is inserted, mainly related to the availability of renewable raw materials as well as the capacity of the installed refining hardware to process the renewable streams.

One of the most common processing routes is the utilization of vegetable or animal oils in the feedstock of conversion or treating units to produce high-quality fuels and petrochemicals. The renewable raw material can be directly processed together with fossil streams in conversion units like fluid catalytic cracking (FCC) to produce transportation fuels and olefins.

The use of renewable streams can also be applied as a feed stream of hydrotreating units, aiming to produce high-quality fuels like diesel and jet fuel.

Some refiners and technology licensors have developed process technologies that make possible the higher synergy of renewables with the conventional refining industry.

In the petrochemical sector, the production of petrochemical intermediates has also adopted renewable processing routes such as ethanol to produce ethylene. Some companies have applied ethylene production through ethanol dehydration since 2010, and some technology licensors have developed processing routes also dedicated to producing ethylene from ethanol.

Renewable streams have a great number of unsaturation and oxygen in their molecules, which lead to high heat release rates and high hydrogen consumption. This fact leads to the necessity of higher capacity of heat removal from hydrotreating reactors aiming to avoid damage to the catalysts. The main chemical reactions associated with the renewable streams hydrotreating process can be represented below:

$$\text{R-CH} = \text{CH}_2 + \text{H}_2 \rightarrow \text{R-CH}_2\text{-CH}_3 \ (\text{Olefins Saturation})$$

$$\text{R-OH} + \text{H}_2 \rightarrow \text{R-H} + \text{H}_2\text{O} \ (\text{Hydrodeoxygenation})$$

where R represents a hydrocarbon.

These characteristics lead to the necessity of higher hydrogen production capacity by the refiners as well as quenching systems of hydrotreating reactors that are more robust or, in some cases, the reduction of processing capacity to absorb the renewable streams. At this point, it's important to consider a viability analysis related to the use of renewables in the crude oil refineries once the higher necessity of hydrogen generation implies higher CO_2 emissions through the natural gas reforming process that is the most applied process to produce hydrogen on a commercial scale.

$$CH_4 + H_2O = CO + 3H_2 \quad \text{(Steam Reforming Reaction–Endothermic)}$$

$$CO + H_2O = CO_2 + H_2 \quad \text{(Shift Reaction–Exothermic)}$$

Despite the concern related to the CO_2 emissions due to hydrogen production, there are some cleaner hydrogen production routes that present attractive alternatives to the downstream players such as the steam reforming of biomethane, reverse water gas shift (RWGS), and the electrolysis process.

These characteristics lead to the necessity of higher hydrogen production capacity by the refiners as well as quenching systems of hydrotreating reactors that are more robust or, in some cases, the reduction of processing capacity to absorb the renewable streams.

This fact leads some technology licensors to dedicate their efforts to looking for alternative routes for hydrogen production on a large scale in a more sustainable manner. Some alternatives pointed out can offer promising advantages:

- Natural gas steam reforming with carbon capture: The carbon capture technology and cost can be limiting factors among refiners.
- Natural gas steam reforming applying biogas: The main difficulty in this alternative is a reliable source of biogas as well as their cost.
- Reverse water gas shift reaction ($CO_2 = H_2 + CO$): One of the most attractive technologies, mainly to produce renewable syngas.
- Electrolysis: The technology is one of the most promising in the near future.

As aforementioned, hydrogen is a key enabler to the future of the downstream industry, and the development of renewable sources of hydrogen is fundamental to the success of the efforts to the energy transition to a lower carbon profile. In the current scenario, the best alternative to refiners is to optimize the hydrogen consumption, minimizing the operating costs and CO_2 emissions.

Another challenge associated with renewable processing is the cold start characteristics of the derivatives, mainly diesel and jet fuel. The renewable feed streams produce highly paraffinic derivatives after the hydrotreating step. In this sense, the final derivative tends to show a higher cloud point, which can be a severe restriction in colder markets such as the northern hemisphere.

In these markets, refiners tend to apply catalytic beds containing dewaxing catalysts (ZSM-5) in their hydrotreating units or cloud point depressor additives, which can raise the operation costs.

Recently, the literature has been classifying the hydrogen production routes into four classes, which are as follows (based on IEA data from 2019):

1. Brown Route: Hydrogen production from coal gasification without carbon abatement system (CCS). This route presents a higher emission rate of greenhouse gases (19 t CO_2/t H_2) and an average production cost of US$ 1.2 to 2.1 per kg H_2.
2. Gray Route: This is the conventional hydrogen production route adopted by most of the refiners, which applies steam reforming of natural gas without CCS. This route still presents high emissions of greenhouse gases (11 t CO_2/t H_2) and an average production cost of US$ 1.0 to 2.1 per kg H_2.
3. Blue Route: This route encompasses the conventional steam reforming of natural gas with a CO_2 abatement system. In this case, the CO_2 emissions are drastically reduced (0.2 t CO_2/t H_2), but the average production cost reaches US$ 1.5 to 2.9 per kg H_2.
4. Green Route: As presented above, the green route is based on electrolysis through renewable electricity. In this case, it's possible to reach zero CO_2 emissions, but the average production cost is still considered high (US$ 3.0 to 7.5 per kg H_2).

The technology development and scale-up gains tend to reduce the production costs of cleaner routes over the next few years. Currently, the best alternative to refiners is to optimize the hydrogen consumption to keep under control the operating costs as well as control the emissions of greenhouse gases.

The energy transition is not a question of choice to the players of the downstream industry; it's a demand from the society and a survival question in the middle term. Decarbonization of the energetic matrix requires even more flexibility and agility by refiners aiming to keep and improve their refining margins in the scenario of reduction in the transportation fuels demand and growing demand by petrochemicals; however, as aforementioned, there are available processing technologies capable of allowing the coprocessing of renewables and fossil feed streams in crude oil refineries, reducing the environmental impact of downstream industry.

Nowadays, it is still difficult to imagine a global energetic matrix free of fossil transportation fuels, especially in developing economies, and raising the participation of renewable raw materials in crude oil refineries can be an attractive strategy. The Brazilian case reinforces that, even in nations with great demand for transportation fuels, biofuels can develop a fundamental role in the energetic matrix.

Despite the recent forecasts indicating a falling demand for transportation fuels and a growing demand for petrochemicals, transportation fuels are still fundamental to sustaining the economic development of nations, especially in developing economies. This fact reinforces, even more, the necessity to reduce the carbon emissions in the crude oil processing chain, and biofuels can play a fundamental role in the achievement of this goal in the downstream industry. Although the environmental benefits are important, it's important to consider that the growing market of biofuels can bring some side effects, like the shortage of renewable feedstock and, in extreme cases, competition between fuels and food production.

The growing participation of biofuels in the energetic matrix also calls the players of the downstream industry to action in order to prepare their refining hardware to increase the rate of renewables coprocessing, aiming to keep and enlarge their participation in the transportation fuels market. At the same time, this scenario can act as a driver for closer integration between refining and petrochemical assets in order to ensure added value to the crude oil through petrochemicals, taking into account the increasing demand destruction of fossil transportation fuels caused by biofuels.

Despite the advantages of biofuels and renewables coprocessing in crude oil refineries, it's fundamental to understand that real decarbonization is only possible through renewable hydrogen production sources as well as efficient carbon capture technologies.

Considering the challenges described above, the downstream players will demand increasing reliability and flexibility of their refining assets, aiming to deal with new raw materials in their processing plants (renewables) as well as to produce and store new derivatives with higher added value and different characteristics. Under this context, the stockpiling of assets and operations will demand even more attention from the refiners to ensure competitive operations.

BIBLIOGRAPHY

Advances in Catalysis for Plastic Conversion to Hydrocarbons – The Catalyst Group (TCGR), 2021.

Avery, C.; Strohm, J. – *FCC Pathways to Co-processing*, PTQ Magazine, 2021.

IEA Bioenergy. *Progress in Commercialization in Biojet/Sustainable Aviation Fuels (SAF): Technologies, Potential, and Challenges* – Technical Report, 2021.

Brazilian Petroleum Agency (ANP). *Brazilian Statistical Yearbook*, 2022.

Chang, R.J. *Crude Oil to Chemicals – Industry Developments and Strategic Implications*, Presented at *Global Refining & Petrochemicals Congress* (Houston, USA), 2018.

Cui, K. – *Why Crude to Chemicals is the Obvious Way Forward.* Wood Mackenzie, 2019.

Energy Research Company (EPE). *Analysis of the Biofuels Conjuncture* – Technical Report, 2020.

Energy Transitions Commission (ETC) – *Making the Hydrogen Economy Possible: Accelerating Clean Hydrogen in an Electrified Economy*, 2021.

Gary, J. H.; Handwerk, G. E. *Petroleum Refining – Technology and Economics.* 4th ed. Marcel Dekker, 2001.

Gelder, A. *Refinery-Petrochemical Integration Disrupts Gas-Based Cracker Feedstock Advantage*, Wood Mackenzie, 2023.

Hilbert, T.; Kalyanaraman, M.; Novak, B.; Gatt, J.; Gooding, B.; McCarthy, S. *Maximising Premium Distillate by Catalytic Dewaxing*, 2011.

IEA Renewables: *Analysis and forecast to 2027*, 2022.

International Energy Agency (IEA). *Oil Outlook*, 2023.

Kim, W.C.; Mauborge, R. *Blue Ocean Strategy*. Harvard Business Review, 2004.

Lindfors, L. P. *High Quality Transportation Fuels from Renewable Feedstock*, Neste Oil Company, 2010.

Porter, M.E. *The Five Competitive Forces that Shape Strategy*. Harvard Business Review, 1979.

Reinventing the Refinery through the Energy Transition and Refining-Petrochemical Integration. IHS Markit, 2020.

Robinson, P.R.; Hsu, C.S. *Handbook of Petroleum Technology*. 1st ed. Springer, 2017.

Rogers, D.L. *The Digital Transformation Playbook: Rethink your Business for the Digital Age*. 1st ed. Columbia University Press, 2016.

Sarin, A.K. *Integrating Refinery with Petrochemicals: Advanced Technological Solutions for Synergy and Improved Profitability* – Presented at *Global Refining & Petrochemicals Congress* (Mumbai, India), 2017.

Silva, M. W. *More Petrochemicals with Less Capital Spending*. PTQ Magazine, 2020.

Vu, T.; Ritchie, J. *Naphtha Complex Optimization for Petrochemical Production*, UOP Company, 2019.

Zhu, F.; Hoehn, R.; Thakkar, V.; Yuh, E. *Hydroprocessing for Clean Energy – Design, Operation, and Optimization*. 1st ed. Wiley Press, 2017.

2 Overview of Crude Oil Refining Processes

2.1 CRUDE OIL DISTILLATION UNITS

The crude oil distillation unit defines the processing capacity of the refinery, and normally, the other process units are sized based on their yields. Figure 2.1 shows a basic process flow diagram for a typical atmospheric crude distillation unit.

The crude oil is pumped from the storage tanks and preheated by hot products that leave the unit in the heat exchanger battery, and then the crude oil stream receives an injection of water aimed to assist the desalting process; this process is necessary to remove the salts dissolved in the petroleum to avoid severe corrosion problems in the process equipment. The desalting process involves the application of an electrical field to the mixture of crude oil and water, which aims to raise the water droplets dispersed in the oil phase and accelerate the decanting; as the salt's solubility is higher in the aqueous phase, a major part of the salts is removed in the aqueous phase

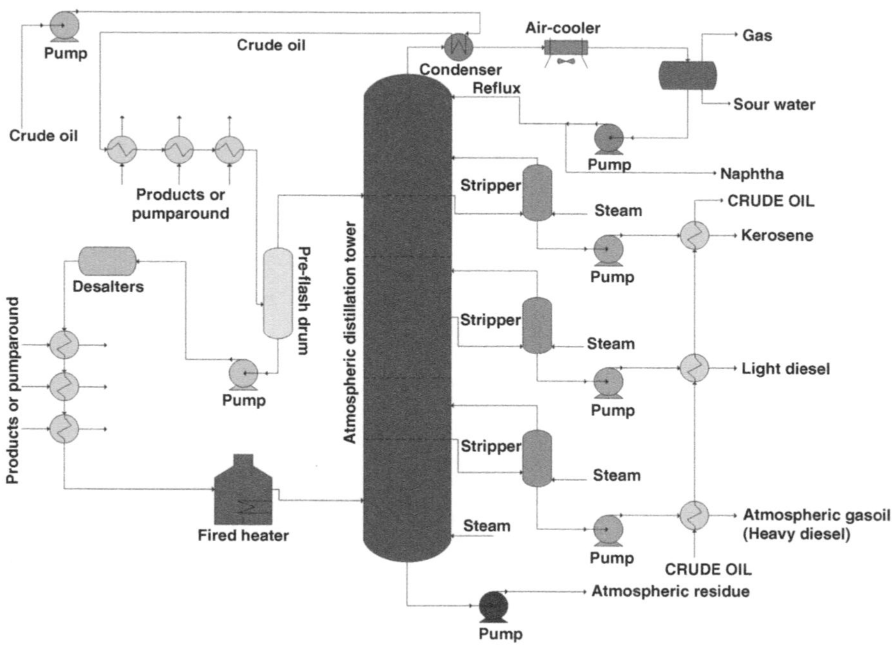

FIGURE 2.1 Process flow diagram for a typical atmospheric crude oil distillation unit.

DOI: 10.1201/9781003472216-2

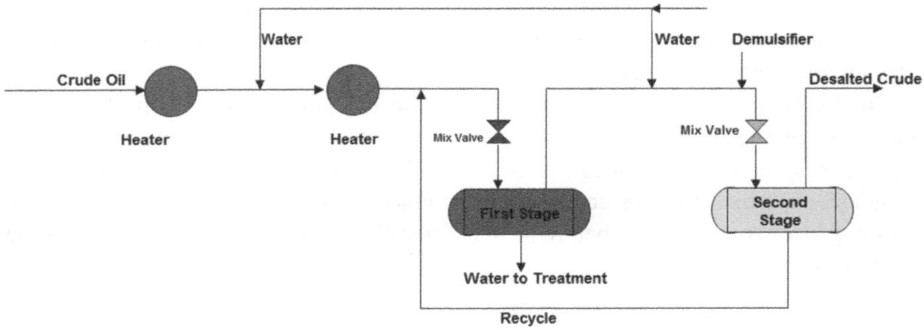

FIGURE 2.2 Crude oil desalting process with two separation stage.

effluent from the desalter, called brine. Normally, the petroleum desalting process is carried out at temperatures between 120 and 160°C. Higher temperatures raise the conductivity of the oil phase and prejudice the phase separation, and this can lead to dragging oil to the brine and result in process inefficiency.

The desalting process involves the mixture of crude oil with water aiming at the dissolution of the salts, considering the higher solubility of these compounds in the aqueous phase. Figure 2.2 depicts a typical desalting process with two separation stages; the salt content in de-desalted crude is normally controlled below 3.0 ppm (as NaCl).

In the desalter exit, the desalted oil is heated again by hot products or pumped around and fed into a flash drum. In this equipment, the lighter fractions are separated and sent directly to the atmospheric tower; the main role of this vessel is reducing the thermal duty needed in the furnace. Following, the stream from the bottom of the flash vessel is heated in the fired heater to temperatures close to 350–400°C (depending on the crude oil to be processed) and is fed to the atmospheric tower, where the crude oil is fractionated according to the distillation range.

At the exit of the atmospheric tower, the products are rectified with steam with the aim of removing the lighter components.

The gaseous fraction is normally directed to the LPG (C_3–C_4) pool of the refinery and the fuel gas system (C_1–C_2), which will feed the furnaces and boilers. The light naphtha is normally commercialized as a petrochemical intermediate or is directed to the gasoline pool of the refining complex. The heavy naphtha can be sent to the gasoline pool, and in some cases, this stream can be added to the diesel pool since it does not compromise the specification requirements of this product (cetane number, density, and flash point). Kerosene is normally commercialized as jet fuel, while the atmospheric residue is sent to the vacuum distillation tower. In some refining schemes, it's possible to send this stream directly to the residue fluid catalytic process unit (RFCC). In this case, the content of contaminant (mainly metals) in the residue needs to be very low to protect the catalyst of the cracking unit.

Nowadays, facing the necessity to reduce the environmental impact of the fossil fuels associated with the restrictive legislations, it is difficult for straight-run products to be commercialized directly. The streams are normally directed to the hydrotreating units, which aim to reduce the contaminants content (sulfur, nitrogen, etc.) before being marketed.

In distillation units with higher processing capacity, normally the flash drum upstream of the atmospheric tower is substituted by a pre-fractionation tower. In these cases, the main advantage is the possibility of reducing the atmospheric tower dimensions, which implies cost reductions associated with the unit implementation and improves the hydraulic behavior in the distillation tower, consequently with better fractionation.

2.1.1 Vacuum Distillation Section

The bottom stream of the atmospheric column (atmospheric residue) still contains recoverable products capable of being converted into high-added-value derivatives; however, under the process conditions of the atmospheric unit, the additional heating led to thermal cracking and coke deposition.

Aiming to minimize this effect, the atmospheric residue is pumped to the vacuum distillation column, where the pressure reduction leads to a reduction in the boiling point of the heavy fractions, allowing the recovery while minimizing the thermal cracking process.

The vacuum generated in the column can be humid, semi-humid, or dry. A humid vacuum occurs when steam injection in the fired heater is applied and in the column aiming to reduce the partial pressure of the hydrocarbons, improving the recovery, while in the semi-humid vacuum, the steam is injected only in the fired heater, minimizing the residence time and reducing the coke deposition. The dry vacuum does not involve the steam injection; in this case, it is possible to achieve pressures between 20 and 8 mmHg while in the humid vacuum, the column operates under pressures varying between 40 and 80 mmHg. However, it's possible to achieve comparable yields through the injection of stripping steam.

As shown in Figure 2.3, the traditional arrangement of vacuum units presents two sides drawn: heavy and light gasoil. These streams are normally directed to conversion units such as hydrocracking or fluid catalytic cracking (FCC), according to the adopted refining scheme. The fractionating quality achieved in the crude oil vacuum distillation column has a direct impact on the reliability and conversion units operation lifecycle. Once this step is controlled, the metal content and the residual carbon (CCR) concentration in the feedstock to these processes are high values of these parameters that lead to a quick catalyst deactivation, raising operational costs and reducing profitability.

Some refiners include additional side withdraws in the vacuum distillation column. When the objective is to maximize diesel production, it's possible to add a withdrawal of a stream lighter than light vacuum gasoil that can be directly added to the diesel pool or after hydrotreating, according to the sulfur content in the processed crude oil. When crude oil presents high metal content, it's possible to include a withdrawal of a fraction heavier than the heavy gasoil called residual gasoil or slop cut. This additional cut concentrates the metals in this stream and reduces the residual

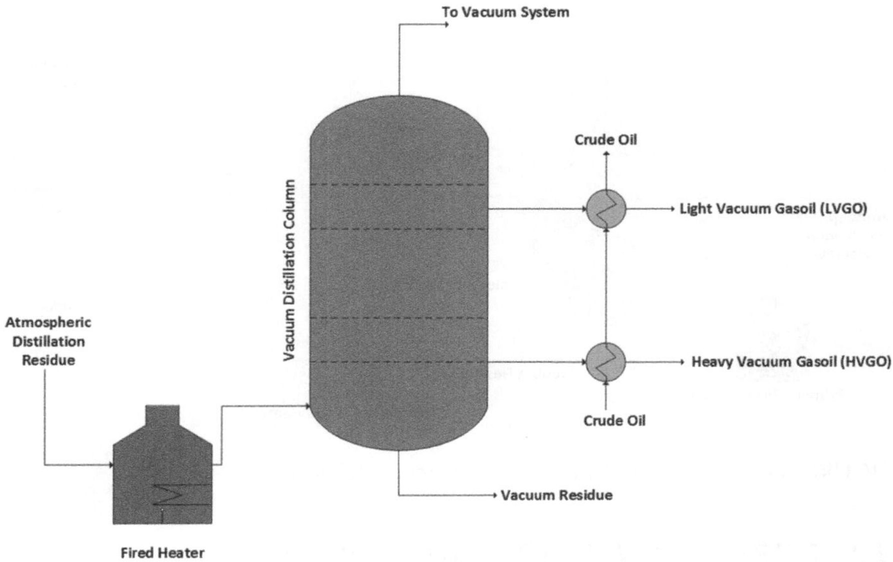

FIGURE 2.3 Schematic process flow diagram for vacuum distillation.

carbon in the heavy gasoil, minimizing the deactivation process of the conversion processes catalysts as aforementioned. The vacuum residue is normally directed to produce asphalt and fuel oils; however, in most modern refineries, this stream is sent to bottom barrel units as delayed coking and solvent deasphalting to produce higher-value products.

According to the refining scheme, the installation of vacuum distillation units can be dispensed with. Refiners that rely on residue fluid catalytic cracking units (RFCC) can send the atmospheric residue directly to the feed stream of these units; however, it's necessary to control the content of the contaminants (metals, sulfur, nitrogen, etc.) and residual carbon (CCR) aiming to protect the catalyst. This fact restricts the crude oil slate that can be processed, reducing the refiner's operational flexibility. On the other hand, in refineries that process extra heavy crudes, normally the crude oil distillation unit is restricted to the vacuum unit once the yields of the atmospheric column are very low and the coking risk is very high.

In refineries optimized to produce lubricants, the distillation process is modified based on the paraffinic characteristics of the crude oil processed, mainly the vacuum distillation step. The necessity to separate the lubricant fractions requires higher fractionation quality in the column, and some configurations rely on two columns, as presented in Figure 2.4.

The distillation unit design is strongly dependent on the characteristics of the crude oil that will be processed by the refinery. For extra-heavy oils, normally, the crude is fed directly to the vacuum column. The design is generally defined based on a limited crude oil range that can be processed in the hardware (contaminant content, API grade, etc.).

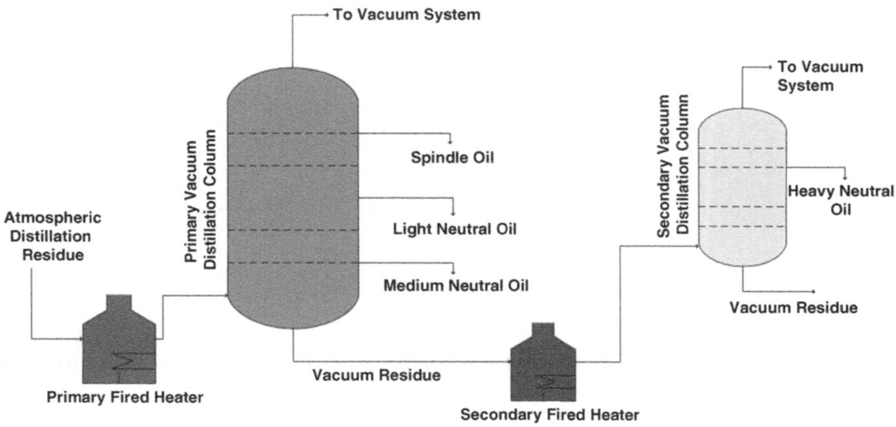

FIGURE 2.4 Vacuum distillation process to produce lubricants.

2.2 THERMAL CONVERSION PROCESSES

The refining processes known as thermal processes are conversion routes which do not apply catalysts to produce the desired reactions, normally cracking reactions which convert larger molecules into higher added value molecules and derivatives.

Due to their characteristics, the thermal cracking processes tend to produce chemically unstable derivatives like olefins due to the absence of hydrogen and the reaction environment, leading to the necessity of a downstream treatment of the intermediate streams before the application into final derivatives.

2.2.1 Visbreaking Process

The search for technologies capable of raising the profitability in oil derivatives production has followed the refining industry since its inception. Over the years, the economic and technological development of nations increased the demand for derivatives drastically, whether for heating, energy production, or transportation fuels.

One of the first technologies applied in the attempt to increase profitability and the yield of the crude oil refining processes was the visbreaking technology.

The visbreaking process consists of thermal cracking under mild conditions and low residence time, aiming to minimize coke production during the feed stream heating step. Figure 2.5 shows a basic process flow diagram for a typical visbreaking process unit.

The visbreaking process aims to reduce the residual oil viscosity destined to be commercialized as fuel oil, avoiding high-added-value streams as dilutants (diesel or gas oil) for this purpose.

Typical feedstock to visbreaking units is the vacuum residue from distillation units or, in some cases, reduced crude oil (after the atmospheric distillation step). The feed stream is heated by the fractionating bottom product before being admitted into the fired heater where the thermal cracking reactions are carried out. In the furnace

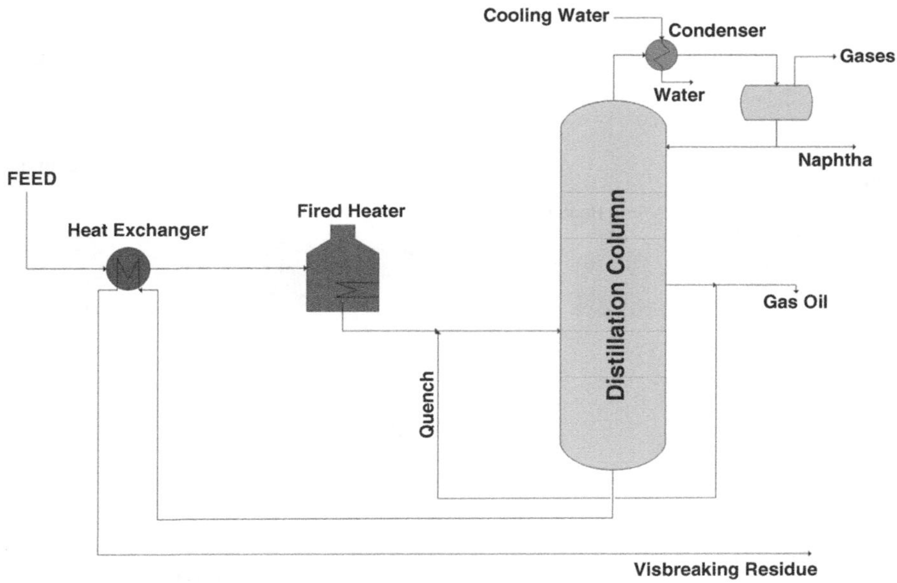

FIGURE 2.5 Typical arrangement for visbreaking process units.

exit, the streams are quenched with gas oil injection, which aims to reduce temperature and stop the thermal cracking reactions before entering the distillation column.

In the distillation column, the products are separated, and the light products and LPG are withdrawn at the top, and the naphtha is directed to the refinery gasoline pool. A gas oil stream is withdrawn and, normally, is directed to compose the feed stream for fluid catalytic cracking unit (FCCU) or hydrocracking process, according to the refining scheme adopted by the refiner. The residual stream from the visbreaking process is directed to fuel oil or asphalt markets, according to the market attended by the refinery.

The process presented in Figure 2.5 is called furnace visbreaking once the thermal cracking reactions occur in the furnace coils.

Another process arrangement for visbreaking process units is shown in Figure 2.6, called soaker visbreaking.

The main difference between these projects' designs is the presence of a vessel after the furnace where a major part of thermal cracking reactions occurs. Called a soaker drum, this equipment is applied to enlarge the residence time, which allows operating the furnace at a lower temperature, consequently with a lower probability of coke production; another advantage is the lower energy consumption in the process.

Another process arrangement for visbreaking units that can be applied is the process where the visbreaking residue is fed to a vacuum distillation column. In these cases, the advantage is a higher gas-oil yield in the process.

The principal process variables for visbreaking units are the furnace temperature, which can vary between 480 and 500°C for the furnace visbreaking and 440 and

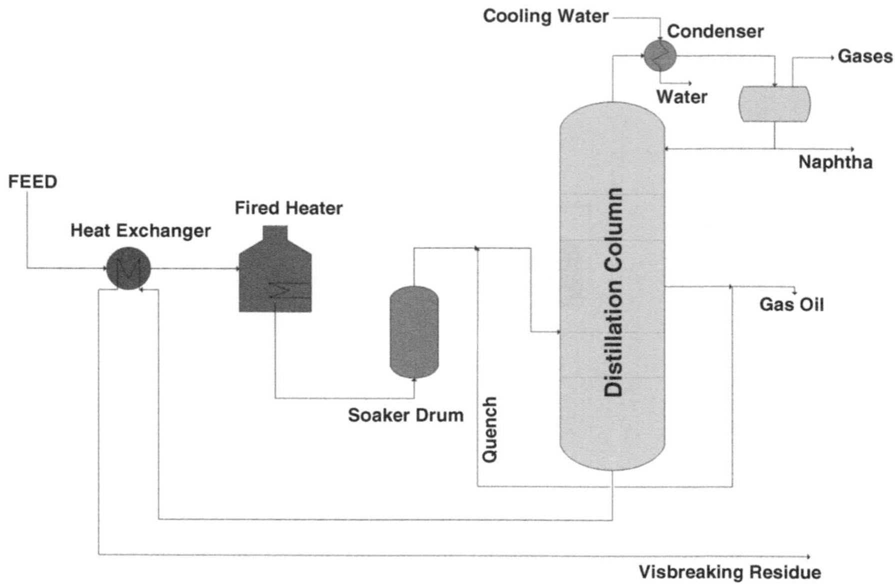

FIGURE 2.6 Visbreaking process with soaker drum.

460°C for the soaker drum visbreaking, and the feed stream pressure can vary from 5 to 35 bar, according to the design concept of the processing unit. The feed stream temperature is normally higher than 300°C. Another relevant process variable is the quench flow rate, applied to interrupt the cracking reactions and achieve an adequate temperature for the fractionating step.

The visbreaking process presents a relatively high operational cost, and once its products are produced from thermal cracking, it needs high-cost treatment to be commercialized (normally hydrotreating), which reduces the process's profitability. Furthermore, the process shows a reduced yield (naphtha and gas oil yield between 15 and 25%). The use of visbreaking units was disseminated in the past; however, after the consolidation of vacuum distillation technology, the process fell into disuse by the refiners.

Nowadays, only specific refining schemes apply visbreaking units; an example is the use of this technology together with delayed coking units aiming to produce high-quality coke (needle or anode grade).

Some technology licensors developed research aiming to raise the competitiveness of the visbreaking process as the SHELL SOAKER VISBRAKING™ technology, licensed by the companies Shell and Lummus. The companies Axens, UOP/Foster Wheeler, and KBR have visbreaking processes in their technology portfolio solutions to the crude oil refining industry.

Despite being an outdated process, as aforementioned, the visbreaking technology can be economically attractive in some scenarios and, from the technological point of view, is still a very interesting process.

2.2.2 Delayed Coking Process

One of the most applied technologies by the refiners is the delayed coking process. Delayed coking employs the thermal cracking concept under controlled conditions to produce light and middle streams (LPG, naphtha, and gas oils) from residual streams which would normally be used as diluents in fuel oil production.

The typical feed stream for delayed coking units is the residue from the vacuum distillation process that contains the heavier fractions of processed crude oil; however, streams like decanted oil from the FCC unit and asphaltic residue produced in solvent deasphalting can compose the feed stream to the delayed coking unit, depending upon the refining scheme adopted by the refiner. Another possibility is to send the residue from atmospheric distillation directly to the delayed coking unit. In this case, the unit design is quite modified, demanding greater robustness of the fractionating and gas compression section.

Due to the thermal cracking characteristics (low availability of hydrogen during the reactions), the streams produced by the delayed coking units have a high concentration of olefinic compounds, which are chemically unstable. Furthermore, due to the processing of residual streams that have high contaminant content, such as nitrogen, sulfur, and metals, the refiners that apply delayed coking units need high hydrotreating capacity to convert these streams into added-value products that meet the contaminant levels according to the environmental regulation.

Figure 2.7 presents the process flow scheme for a typical delayed coking unit.

The feed stream is fed into the bottom of the main fractionating tower, where it is mixed with the heavier fraction of the thermal cracking products and then sent to the fired heater, where thermal cracking reactions are initiated. The reaction conditions are controlled so that the reactions are completed in the coke drums. The residence

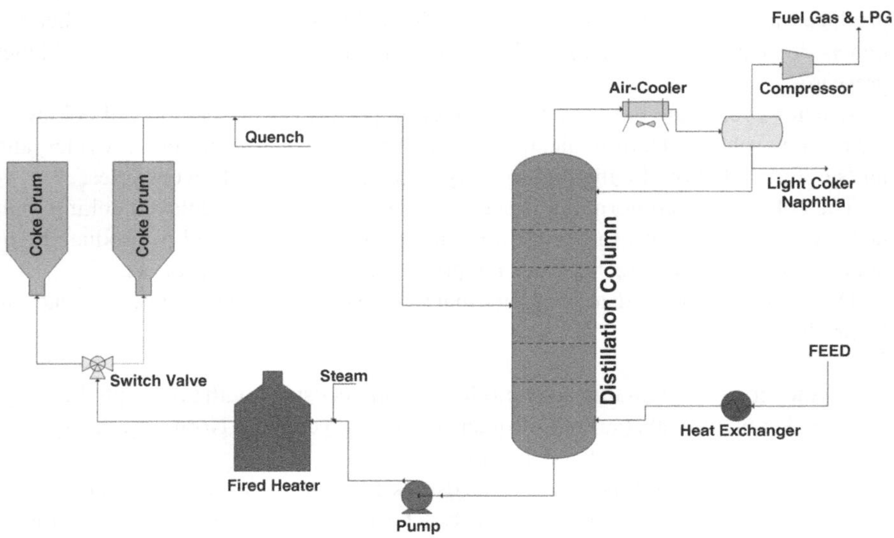

FIGURE 2.7 Typical arrangement for delayed coking unit.

time in the fired heater must be the lowest possible to minimize the coke precipitation in the fired heater tubes. A way of minimizing the coke formation in the walls of tubes is the steam injection that raises the velocity and consequently reduces the residence time.

After the fired heater, the feed stream is sent to the coke drum or reactor, where the thermal reactions are completed, and the coke is deposited. The thermal cracking products are removed from the top of the reactor and receive an injection of quench with a cold process stream (normally heavy or middle gas oil) and are directed to the main fractionators where the products are separated. The coke deposited in the reactor is removed through a cut with water under high pressure (about 250 bars).

Delayed coking is a process that occurs in batches. In order to make it a semi-continuous process, pairs of reactors are always employed, and each two reactors is applied one fired heater. When one reactor is under reaction, the other is in the decoking step and so on. The delayed coking process occurs in cycles that can vary from 14 to 24 hours.

The main operational variables of the delayed coking unit are the recycling ratio, which is the quantity of the total feed stream which corresponds to the heavier fraction of the reaction products that are mixed with the fresh feed; reactor temperature, normally considered in the top of the coke drum; pressure in the top of the reactor; and the time of the reactor cycle.

The recycle ratio varies normally between 5 and 10% (to units dedicated to producing fuels), and the refiners seek to operate the unit with the lowest recycle ratio possible in order to maximize the capacity of the plant in processing residual streams. The reactor temperature is close to 430°C and is linked with the fired heater temperature. Throughout the thermal cracking reactions, the temperature falls due to the endothermic characteristics of the reactions.

The pressure in the reactor can vary between 1 and 3.5 bars. In units optimized to produce fuels, the variable is maintained at lower levels; on the other hand, when the unit is dedicated to producing high-quality coke, the unit is operated under higher pressures.

Reactor cycle time is linked to the function performed by the delayed coking in the refining scheme. Units dedicated to producing fuels operate at shorter cycles, and units optimized to producing high-quality coke operate under longer cycles.

The coke produced normally is seen as a by-product of the delayed coking unit; however, in some cases, the delayed coking process is optimized to produce high-quality coke, and the coke becomes the principal product of the process.

Depending on the feedstock quality that will be processed, three types of coke can be produced:

- Shot coke: Poor quality coke produced from feedstock with high asphaltenes and contaminants (sulfur, nitrogen, and metals) content. Normally, this type of coke is commercialized as fuel.
- Sponge coke: In this case, the feedstock has a lower asphaltenes and contaminants content, and the coke can be directed to raw material for the anodes production process in the aluminum industry.

- Needle Coke: The production of this type of coke requires the processing of feedstock with high aromatic content (e.g., decanted oil from FCC), and these products are sent as raw material to produce anodes for the steel industry.

As mentioned above, the production of high-quality coke requires quality control of the feed stream that will be processed. In most cases, the refiners choose to install delayed coking units focusing on the production of middle and light distillates. Therefore, the unit optimization to produce needle coke occurs only in specific cases.

Figure 2.8 shows a delayed coking main fractionator scheme with the principal process streams.

The heavy gas oil stream is normally directed to the FCCU or can be utilized as fuel oil. In refining schemes that have deep hydrocracking units, this stream can be used as feedstock to the unit. The sending of this stream to the FCCU needs to be controlled to avoid the premature deactivation of the catalyst, the face of the high level of contaminants, mainly nitrogen and metals.

Middle and light gas oils are normally sent to severe hydrotreating units to compose the diesel pool of the refinery. The heavy coker naphtha can be directed like a feed stream to FCC units. When the flash point specification of diesel is not restricted, this stream can be sent to the diesel pool after a deep hydrotreating process.

The lighter fraction of naphtha can be sent to the gasoline pool of the refinery after hydrotreatment or directed to FCC units. In this case, this stream contributes to

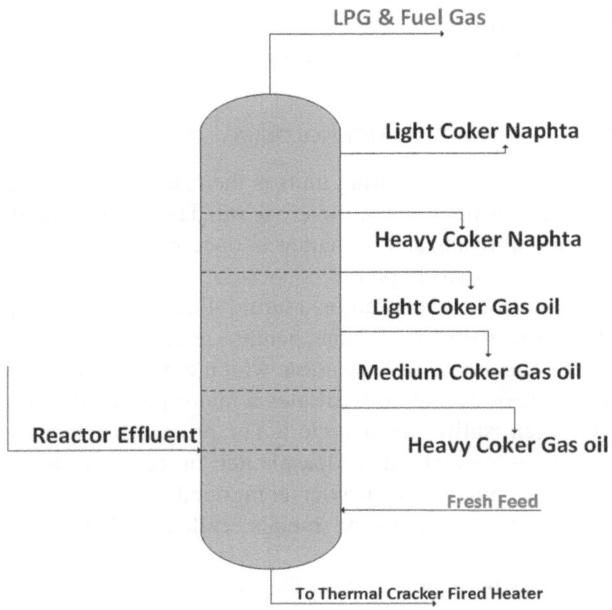

FIGURE 2.8 Main fractionator scheme for a typical delayed coking unit.

raising the LPG production in the FCC unit. In some cases, the light coker naphtha can be sent to catalytic reforming units, aiming to produce high octane gasoline or petrochemical precursors (benzene, toluene, and xylenes).

The overhead products from the main fractionator are still in the gaseous phase and are sent to the gas separation section. The fuel gas is sent to the refinery fuel gas ring, after treatment to remove H_2S, where it will be burned in fired heaters while the LPG is directed to treatment and further commercialization.

Figure 2.9 presents a typical scheme for a gas separation section for a delayed coking unit.

The main licensors for delayed coking technology nowadays are the companies FOSTER WHEELER™, CONOCOPHILLIPS™, LUMMUS™, and KBR™.

Delayed coking technology becomes especially attractive for refiners installed in countries with large heavy and extra-heavy crude oil reserves like Mexico and Venezuela. The use of delayed coking in the refining scheme can minimize the production of low-added-value products like fuel oils and guarantee higher flexibility to the refinery in relation to processed crude oil, minimizing the necessity to acquire light oils.

On the other hand, the delayed coking technology obliges the refiners to the necessity of high hydrotreatment capacity once the streams produced by the unity need a severe treating process before being sent to commercialization. This fact can raise the operational and installation costs.

Another delayed coking disadvantage is the necessity of handling, storage, and commercialization of coke, which normally are not the focus of the refiners. Some variations of coking technology, such as FLUID COKING e FLEXICOKING™, the last licensed by ExxonMobil™ Company, can minimize or eliminate the coke formation during the oil residual streams upgrading.

2.2.3 Solvent Deasphalting Technologies

The typical feedstock for deasphalting units is the residue from vacuum distillation that contains the heavier fractions of the crude oil. The residue stability depends on the equilibrium among resins and asphaltenes, once which the resins solubilize the asphaltenes, keeping a dispersed phase.

The deasphalting process is based on a liquid–liquid extraction operation where light paraffin is applied (propane, butane, pentane, etc.) to promote resin solubilization, inducing the asphaltenes precipitation, which corresponds to the heavier fraction of the vacuum residue and concentrates a major part of the contaminants and heteroatoms (nitrogen, sulfur, metals, etc.). The process produces a heavy stream with low contaminant content called deasphalted oil (extract phase) and a stream poor in a solvent containing the heavier compounds and with high contaminant content, mainly sulfur, nitrogen, and metals, called asphaltic residue (raffinate phase).

Figure 2.10 shows a basic process flow diagram for a typical process deasphalting unit.

The vacuum residue is fed to the extracting tower, where contact with the solvent leads to the saturated compound solubilization. In the sequence, the mixture of

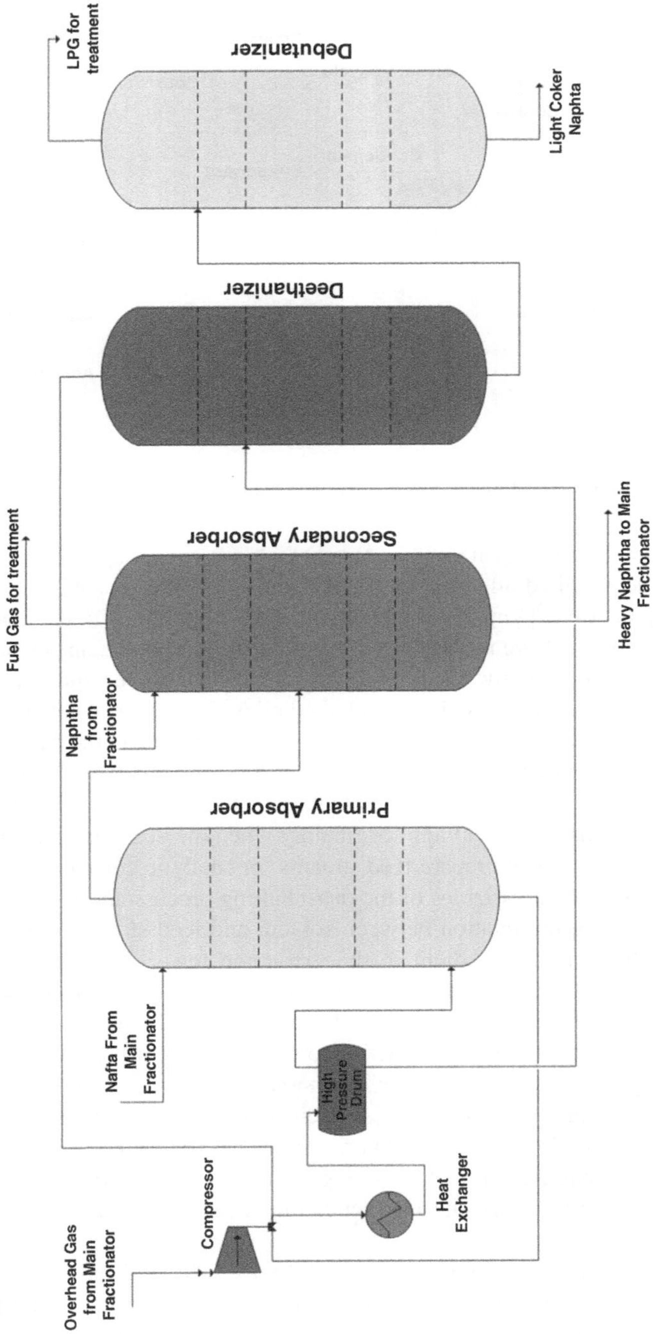

FIGURE 2.9 Basic process flow diagram for a typical gas separation section from delayed coking unit.

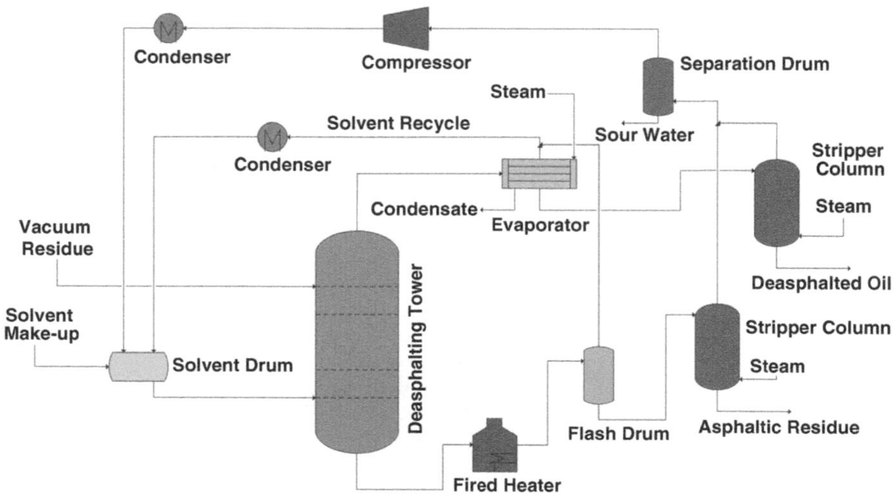

FIGURE 2.10 Typical arrangement for a solvent deasphalting process unit.

solvent/vacuum residue is sent to separation vessels where the separation of asphaltic residue from deasphalted oil occurs, as well as the solvent recovery.

The choice of solvent employed has fundamental importance to the deasphalting process. Solvents that have higher molar mass (higher carbon chain) present higher solvency power and raise the yield of deasphalted oil; however, these solvents are less selective, and the quality of the deasphalted oil is reduced once heavier resins are solubilized, which leads to a higher quantity of residual carbon in the deasphalted oil; consequently, the content of contaminants raises too. As normally, the deasphalting unit aims to minimize the carbon residue, metals, and heteroatoms in the deasphalted oil; propane is the usual solvent applied, mainly when the deasphalting process role in the refining scheme is to prepare feed streams for catalytic conversion processes.

The main operational variables of the deasphalting process are feedstock quality, solvent composition, the relation between solvent and feed stream, extraction temperature, and temperature gradient in the extraction tower. Despite being a very important variable, the extraction pressure is defined in the unit design step and is normally defined as the needed pressure to keep the solvent in the liquid phase. In the propane case, the pressure in the extraction tower is close to 40 bar.

Feedstock quality depends on crude oil characteristics processed by the refinery, as well as the vacuum distillation process. Depending on the fractionation produced in the vacuum distillation unit, the vacuum residue can be heavier or lighter, directly affecting the deasphalting unit yield. Using propane as a solvent, the relation solvent/feed stream is close to 8, and the feed temperature in the extraction tower is close to 70°C.

In refineries focused on fuel production (mainly LPG and gasoline), the deasphalted oil stream is normally sent to the FCCU. In this case, the contaminants content and carbon residue need to be severely controlled to avoid premature deactivation of the catalyst, which is very sensitive to metals and nitrogen. In refineries dedicated to producing middle distillates, the deasphalted oil (DAO) can be directed to hydrocracking units.

When the deasphalting process is installed in refining units dedicated to producing lubricants, the quality of deasphalted oil tends to be superior in view of the fact that the crude oil processed is normally lighter and with lower contaminant content. In this case, the deasphalted oil is directed to an aromatic extraction unit or to hydrotreatment/hydrocracking units. In the last case, the deasphalted oil quality is more critical because of the possibility of premature catalyst deactivation.

The asphaltic residue stream is sent to the fuel oil pool after dilution with lighter compounds (gas oils), or the stream can be used to produce asphalt. Another possibility is to send the asphaltic residue to a delayed coking unit. As the aromatic content in the asphaltic residue is high, the coke produced presents a very good quality.

The principal step in the solvent deasphalting process is the liquid–liquid extraction, which depends strongly on the solvent properties; in this sense, some licensors developed deasphalting processes based on the solvent in supercritical conditions. Above the critical point, the solvent properties are more favorable to the extraction process, mainly solvency power and the vaporization and compression facility, which reduce the power consumer in the process, leading to lower operating costs.

The processes ROSE™ licensed by KBR Company, UOP-DEMEX™ licensed by UOP, and the process SOLVAHL™ licensed by AXENS are examples of deasphalting technologies in supercritical conditions. Figure 2.11 presents a basic process scheme for a typical deasphalting unit under supercritical conditions.

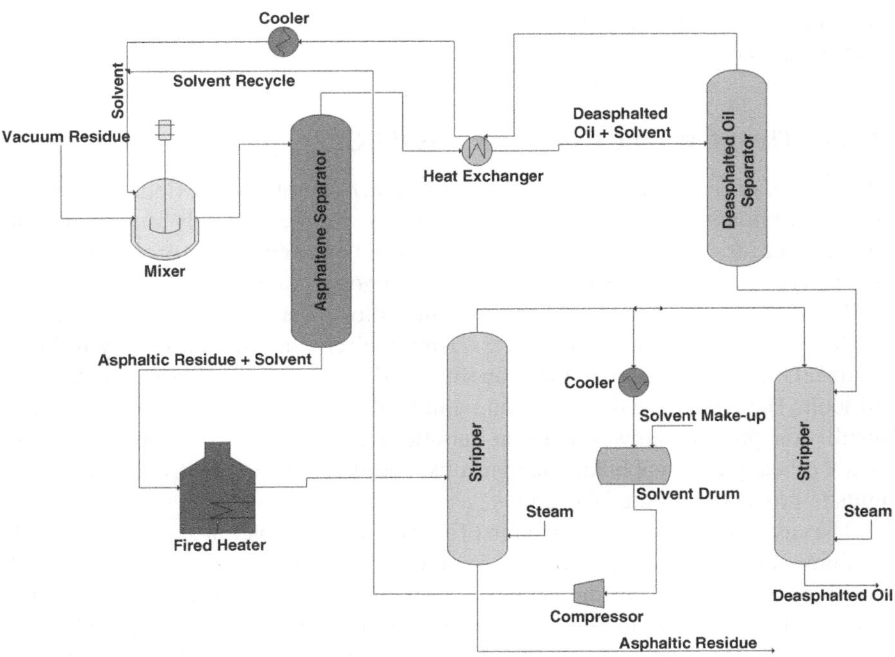

FIGURE 2.11 Typical arrangement to solvent deasphalting unit under supercritical condition.

In addition to the quoted processes, the FOSTER WHEELER Company, in partnership with UOP, developed the process UOP/FW-SDA™, which also applies solvent in supercritical conditions.

As described earlier, the deasphalting process allows adding value to residual streams as vacuum residue and, consequently, raises the refiners' profitability. Furthermore, the process can help in the production of higher quality and cleaner derivatives.

As another residue upgrading technology, the deasphalting process raises the refinery's flexibility regarding the quality of processed crude oil, which can pass to process heavier crude oils that have normally lower costs, and this fact can improve the refining margin.

Currently, deasphalting technology has lost ground in the more modern refining schemes to delayed coking units since these units can process residual streams, producing streams that can be converted into products with high added value (LPG, gasoline, and diesel) without the need for previous feed stream treatment to remove contaminants. However, the products from delayed coking units need hydrotreatment to be commercialized, which significantly raises the operational and installation costs of the refinery.

2.3 CATALYTIC CONVERSION PROCESSES

As indicated by the name, the catalytic conversion processes apply catalysts to promote the conversion reactions. The catalytic processes can be applied to remove contaminants from final or intermediate streams, like the hydrotreating units, or to produce new and higher added value molecules such as hydrocracking, catalytic reforming, and FCCUs.

2.3.1 Fluid Catalytic Cracking Units (FCC)

FCC is one of the main processes that give higher operational flexibility and profitability to refiners. The catalytic cracking process has been widely studied over the last few decades and has become the principal and most employed process dedicated to converting heavy oil fractions into higher economic value streams.

The installation of catalytic cracking units allows the refiners to process heavier crude oils and consequently cheaper, raising the refining margin, mainly in higher crude oil price scenarios or in the geopolitical crises that can make access to light oils difficult. The typical catalytic cracking unit feed stream is gas oils from the vacuum distillation process. However, some variations are found in some refineries, like sending heavy coke naphtha, coke gas oils, and deasphalted oils from deasphalting units to processing in the FCC unit.

The catalyst normally employed in FCCUs is a solid constituted of small particles of alumina (Al_2O_3) and silica (SiO_2) (zeolite). To the catalyst characteristics and the operational conditions in the catalytic cracking process (temperature higher than 500°C), the process is inefficient for cracking aromatic compounds; therefore, the more paraffinic the feed stream, the higher the unit conversion. Figure 2.12 presents a process scheme for a typical FCCU.

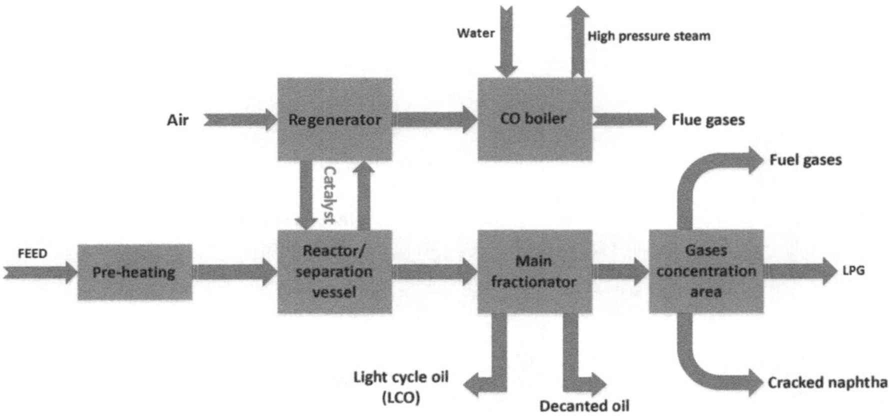

FIGURE 2.12 Schematic process flow for a typical fluid catalytic cracking process unit (FCCU).

In a conventional scheme, the catalyst regeneration process consists of the carbon partial burning deposited over the catalyst, according to the chemical reaction below:

$$C + \frac{1}{2} O_2 \rightarrow CO$$

The carbon monoxide is burned in a boiler capable of generating higher-pressure steam that supplies other process units in the refinery.

The principal operational variables in an FCCU are reaction temperature, normally considered the temperature at the top of the reactor (called a riser), feed stream temperature, feed stream quality (mainly carbon residue), feed stream flow rate, and catalyst quality. Feedstock quality is especially relevant, but this variable is a function of the crude oil processed by the refinery, so it is difficult to change; for example, aromatic feedstocks with high metal content are refractory to cracking and conducting quick catalyst deactivation.

An important variation of the FCC technology is the RFCC unit. In this case, the feedstock to the process is basically residue from the atmospheric distillation column, due to the high carbon residue and contaminants (metals, sulfur, nitrogen, etc.), which are necessary for some adaptations in the unit, like a catalyst with higher resistance to metals and nitrogen and catalyst coolers. Furthermore, it's necessary to apply materials with the most noble metallurgy due to the higher temperatures reached in the catalyst regeneration step (due to the higher coke quantity deposited on the catalyst), which significantly raises the capital investment for the unit installation. Nitrogen is a strong contaminant to the FCC catalyst because it neutralizes the acid sites of the catalyst, which are responsible for the cracking reactions.

When the residue has a high contaminant content, it is common for the feed stream treatment in hydrotreating units to reduce the metals and heteroatom concentration to protect the FCC catalyst.

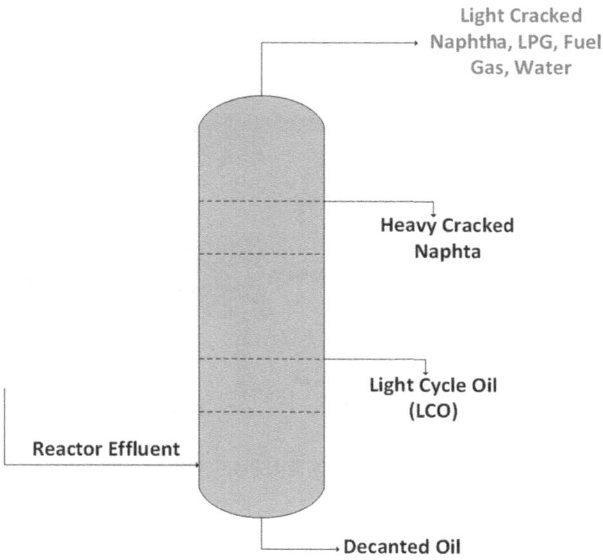

FIGURE 2.13 Main fractionator scheme for a typical fluid catalytic cracking unit.

Typically, the average yield in FCCUs is 55% in volume in cracked naphtha and 30% in LPG. Figure 2.13 presents a scheme for the main fractionator of the FCC unit with the principal product streams.

The decanted oil stream contains heavier products and has high aromatic content. It is common that this product is contaminated with catalyst fines, and normally this stream is directed to use like fuel oil diluent, but in some refineries, this stream can be used to produce black carbon.

Light cycle oil (LCO) has a distillation range close to diesel, and normally this stream is directed to treatment in severe hydrotreating units (due to the high aromaticity); after this treatment, the LCO is sent to the refinery diesel pool.

Heavy cracked naphtha is normally directed to the refinery gasoline pool; however, in scenarios where the objective is to raise the production of middle distillates, this stream can be sent to hydrotreating units for further diesel production.

The overhead products from the main fractionator are still in the gaseous phase and are sent to the gas separation section. The fuel gas is sent to the refinery fuel gas ring after treatment to remove H_2S, where it will be burned in fired heaters while the LPG is directed to treatment (MEROX) and further commercialization. The LPG produced by the FCC unit has a high content of light olefins (mainly propylene), so, in some refineries, the LPG stream is processed in a propylene separation unit to recover the propylene that has higher added value than LPG.

Cracked naphtha is usually sent to the refinery gasoline pool, which is formed by naphtha produced by other process units such as straight-run naphtha and naphtha from the catalytic reforming unit. Due to the production process (deep conversion of residues), the cracked naphtha has high sulfur content, and to attend to the current environmental legislation, this stream needs to be processed to reduce the content of contaminants, mainly sulfur.

The cracked naphtha sulfur removal represents a great technology challenge because it is necessary to remove the sulfur components without molecule saturation that gives a high octane number for gasoline (mainly olefins).

Over the last decades, some technology licensers have developed new processes, aiming to reduce the sulfur content in the cracked naphtha with minimum octane number loss. Some of the main technologies dedicated to this purpose are PRIME G+™ technology from Axens, the OCTAGAIN ™ and SCANfining ™ processes from Exxon Mobil, the S-Zorb™ process from ConocoPhillips, and ISAL™ technology from UOP.

Usually, catalytic cracking units are optimized to aim at the production of fuels (mainly gasoline); however, some process units are optimized to maximize the light olefins production (propylene and ethylene). Process units dedicated to this purpose have his project and operational conditions significantly changed once the process severity is strongly raised in this case.

The reaction temperature reaches 600°C and a higher catalyst circulation rate raises the gas production, which requires a scaling up of the gas separation section. Figure 2.14 presents a typical scheme for a gas separation section for an FCCU.

In several cases, due to the higher heat necessity of the unit, it is advantageous to operate the regenerator with the total combustion of the coke deposited on the catalyst. This arrangement significantly changes the thermal balance of the refinery once it's no longer possible to resort to the steam produced by the CO boiler.

Over the last decades, the FCC technology has been intensively studied, aiming mainly at the development of units capable of producing light olefins (deep catalytic cracking) and processing heavier feedstocks. The main licensers for FCC technology nowadays are companies such as KBR, UOP, STONE and WEBSTER, AXENS, LUMMUS, and FOSTER WHEELER.

Despite the great operational flexibility that FCC technology gives for the refineries, some new projects have dismissed these units in the refining scheme, mainly when the new refinery objective is to maximize middle distillate products (diesel and kerosene) once this is not the focus of the FCCU.

2.4 CATALYTIC REFORMING TECHNOLOGIES

The main objective of the catalytic reforming unit is to produce a stream with a high aromatic hydrocarbon content that can be directed to the gasoline pool or to produce petrochemical intermediates (benzene, toluene, and xylenes) according to the market served by the refiner. Due to the high content of aromatic compounds, the reformate can significantly raise the octane number in the gasoline; in the current scenario, this is a less attractive route.

A typical feedstock to the catalytic reforming unit is the straight-run naphtha; however, in the last decades, due to the necessity to increase the refining margin through the installation of bottom barrel units, hydrotreated coke naphtha stream has been consumed like feedstock in the catalytic reforming unit.

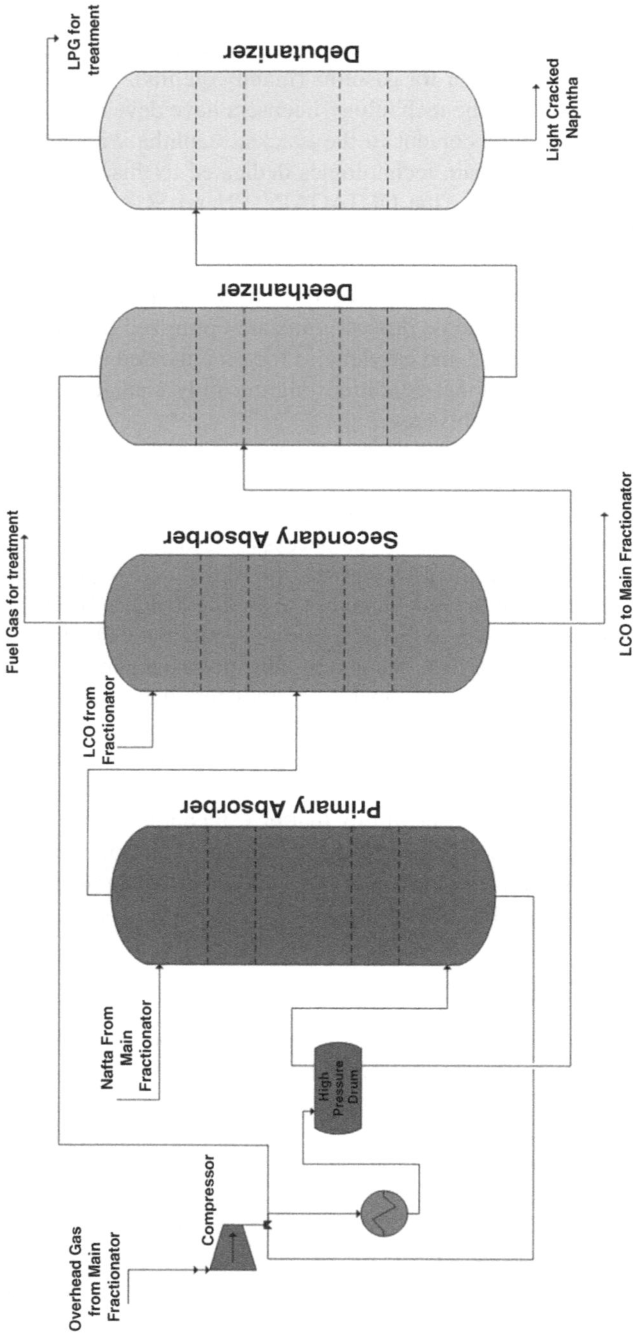

FIGURE 2.14 Basic process flow diagram for a typical gas separation section from FCC unit.

The catalyst generally employed in the catalytic reforming process is based on platinum (Pt) supported by alumina treated with chlorinated compounds to raise the support acidity. This catalyst has bifunctional characteristics once the alumina acid sites are active in molecular restructuring, and the sites of metal are responsible for hydrogenation and dehydrogenation reactions.

The main chemical reactions involved in the catalytic reforming process are:

- Naphthene compounds dehydrogenation.
- Paraffins isomerization.
- Isomerization of naphthene compounds.
- Paraffins dehydrocyclization.

Among the undesired reactions can be cited hydrocracking reactions and dealkylation of aromatic compounds.

Figure 2.15 presents a basic process flow diagram for a typical semi-regenerative catalytic reforming unit.

The naphtha feed stream is blended with recycled hydrogen and heated at a temperature varying from 500 to 550°C before entering the first reactor. As the reactions are strongly endothermic, the temperature falls quickly, so the mixture is heated and sent to the second reactor and so on. The effluent from the last reactor is sent to a separation drum where the phases of liquid and gaseous are separated.

The gaseous stream with high hydrogen content is shared in two process streams: a part is recycled to the process to keep the ratio H_2/feed stream and the other part is sent to a gas purification process plant (normally a pressure swing adsorption unit) to raise the purity of the hydrogen that will be exported to others process plants in the refinery.

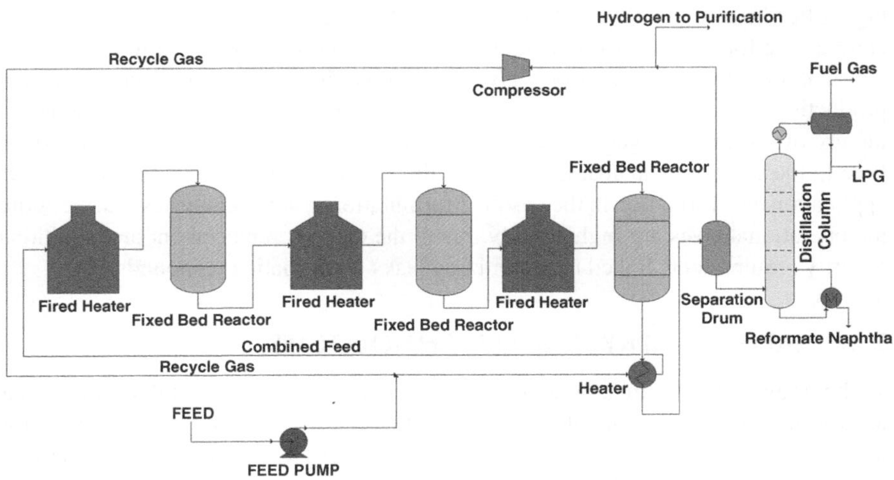

FIGURE 2.15 Typical arrangement of semi-regenerative catalytic reforming process unit.

The liquid fraction obtained in the separation drum is pumped to a distillation column wherein the bottom is produced, the reformate, and in the top drum of the column is produced LPG and fuel gas.

The reformate has a high aromatics content and, according to the market supplied by the refinery, can be directed to the gasoline pool like a booster of octane number, or, when the refinery has aromatics extraction plants, it is possible to produce benzene, toluene, and xylenes in segregated streams, which can be directed to petrochemical process plants. The gas rich in hydrogen produced in the catalytic reforming unit is an important utility for the refinery, mainly when there is a deficit between the hydrogen production capacity and the hydrotreating installed capacity in the refinery. In some cases, the catalytic reforming unit is operated with the principal objective to produce hydrogen.

The main process variable in the catalytic reforming process unit is pressure (3.5–30 bar), which normally is defined in the design step; in other words, the pressure normally is not an operational variable. The temperature can vary from 500 to 550°C. The space velocity can be varied through feed stream flow rate control and the ratio H_2/feed stream that have a direct relation with the quantity of coke deposited on the catalyst during the process. To semi-regenerative units, the ratio of H_2/feed stream can vary from 8 to 10. In units with continuous catalyst regeneration, this variable can be significantly reduced.

Due to the process severity, the high coke deposition rate on the catalyst, and consequently the quick deactivation, there are short operational campaign periods for semi-regenerative units that employ fixed bed reactors.

To solve this problem, some technology licensors developed a catalytic reforming process with continuous catalyst regeneration steps.

In the regeneration section, the catalyst is submitted to processes to burn the coke deposited during the reactions and treated with chlorinated compounds to reactivate the acid function of the catalyst.

Despite the higher capital investment, the rise in the operational campaign and higher flexibility in relation to the feedstock to be processed in the processing unit can compensate for the higher investment in relation to the semi-regenerative process.

Catalytic reforming technology gives great flexibility to the refiners in the gasoline production process; however, in the last decades, there has been a strong restriction on the use of reformate in gasoline due to the control of benzene content in this derivate (due to the carcinogenic characteristics of this compound). This fact has reduced the application of reformation in the gasoline formulation in some countries. Furthermore, the operational costs are high, mainly due to the catalyst replacement and additional security requirements linked to minimizing leaks of aromatic compounds.

2.5 NAPHTHA ALKYLATION TECHNOLOGIES

Technological development associated with restrictive environmental legislation has led to the development of internal combustion engines which need higher performance fuels, lower consumption, and low emissions of environmentally harmful gases (NOx, SOx, etc.).

This scenario presents a growing challenge to the crude oil refining industry and process technology developers, mainly when the maturation of new transport alternatives like electric vehicles and hybrid engines threatens strongly the crude oil derivates market, such as diesel and gasoline.

Gasoline is one of the most consumed crude oil derivates and is normally produced through a mixture of naphtha from different refining process steps. The streams normally involved in the gasoline production process are straight-run naphtha, cracked naphtha, coke naphtha (after hydrotreatment), and reforming naphtha.

One of the main parameters of gasoline quality is the octane number, which is a measure of the combustion quality of this derivate. One of the streams that contribute to raising the octane number is the reformed naphtha, produced in the catalytic reforming unit. However, due to the severe restrictions related to the carcinogenic aromatic emissions, mainly benzene, some refiners have avoided the application of this stream to formulate gasoline, directing the reformed naphtha preferably to petrochemical intermediates production in aromatics complexes.

An alternative to the reforming naphtha is the production of branched hydrocarbons (with high octane number) through the catalytic alkylation process.

The alkylation process involves the reactions between light olefins (C_3–C_5) and isoparaffinic hydrocarbons like isobutane. The reaction product called alkylate is a mixture of branched hydrocarbons with higher molecular weight and higher octane number.

An example of a typical alkylation reaction is represented below:

$$C_4H_{10} + C_3H_6 \rightarrow C_7H_{16}\left(2,3\,\text{Dimethylpentane}\right)$$

The reaction is catalyzed in strongly acidic reaction environments. The acids normally employed in the industrial scale technologies are hydrofluoric acid (HF) and sulfuric acid (H_2SO_4).

The main advantage of the alkylation process is the production of a stream with a high octane number, high chemical stability, and practically free of contaminants such as nitrogen and sulfur. These characteristics turn the alkylate into a component attractive to the gasoline formulation in the automotive and aviation industries.

Alkylation process feed streams are generally obtained from LPG produced in deep conversion units, mainly FCC and delayed coking. The LPG produced in these process units has high olefin content, ideal for the alkylation process. The isobutane stream is normally obtained through the separation of LPG produced in the atmospheric distillation unit, FCC, or delayed coking in deisobutanizer towers.

As aforementioned, the acids generally employed as the homogeneous catalysts in the alkylation process are HF and H_2SO_4. Figure 2.16 presents a process flow diagram of the alkylation process catalyzed by HF.

The feed stream goes through pretreatment (generally molecular sieves or alumina) before being pumped to the reactor. The objective is to remove process contaminants, mainly water, diolefins, sulfur, and nitrogen compounds. Water is especially damaging to the process; once it accelerates piping and equipment corrosion processes, and beyond that, it requires higher HF reposition.

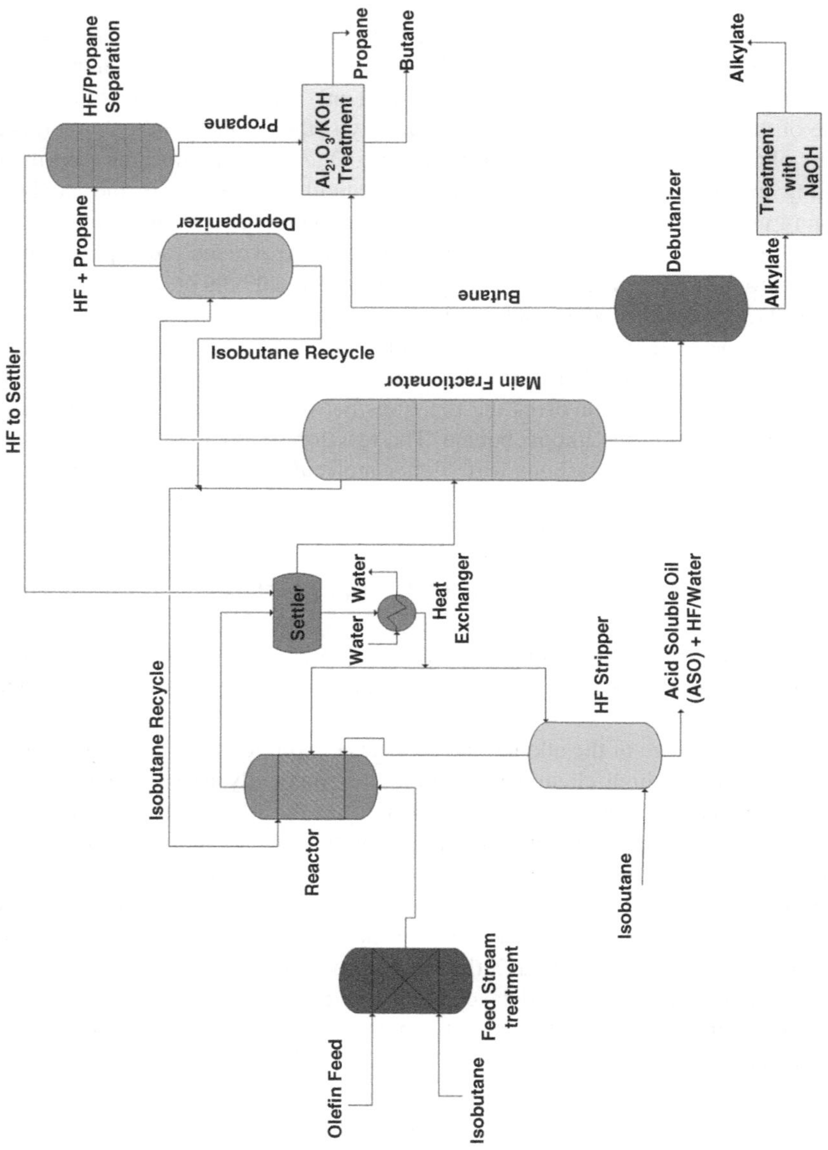

FIGURE 2.16 Typical process flow diagram to catalytic alkylation unit using HF as catalyst.

After pretreatment, the hydrocarbon streams are put in contact with the hydrofluoric acid in the reactor, and the hydrocarbon mixture and HF solution are separated through gravity in a settler vessel. The hydrocarbon phase is sent to the fractionating section while the aqueous phase (containing most of the HF) is cooled and sent back to the reactor. As alkylation reactions are exothermic, the reactor is continuously refrigerated, aiming to keep the reaction in ideal conditions.

A part of hydrofluoric acid is sent to the stripping column, where the acid is stripped with isobutane. The top product is a mixture of HF and isobutene and is sent back to the reactor, while the bottom stream contains an azeotropic mixture of water and HF. Beyond hydrocarbons, this step is responsible for keeping the HF free of contaminants and with adequate concentration in the alkylation process.

After the separation columns, butane and propane streams go to treatment with alumina, aiming to decompose organic fluorides and with KOH to neutralize the remaining acidity. The alkylate stream is treated with NaOH to neutralize the remaining acidity. Currently, the main alkylation technology licensors with HF are companies such as UOP and CONOCO-PHILLIPS.

Alkylate stream is normally directed to the refinery gasoline pool for the production of high-octane automobile gasoline or aviation gasoline; however, in petrochemical plants, this stream can be used as an intermediate to produce ethylbenzene (for styrene production), isopropyl-benzene (to produce phenol and acetone), and dodecyl-benzene used to produce detergents. Propane and butane streams can be sent to the LPG pool of the refinery or commercialized separately.

The alkylation process with sulfuric acid as a catalyst has similarities with the hydrofluoric acid process; however, the sulfuric acid regeneration step is more complex and involves the H_2SO_4 decomposition in SO_2 and SO_3 and the subsequent condensation of concentrated sulfuric acid. This regeneration can be conducted in the processing site or in an external process plant; consequently, the sulfuric acid consumption in the process is much higher than HF; furthermore, the solubility of sulfuric acid in hydrocarbons is lower, requiring greater agitation to maintain the contact between the phases in ideal conditions to the process.

The alkylation technologies with sulfuric acid most applied on an industrial scale are the processes STRATCO Effluent Refrigerated Alkylation Process™, licensed by STRATCO Engineering Company, and EXXON MOBIL Cascade Autorefrigerated Process™, licensed by EXXON MOBIL Company. Figure 2.17 shows a simplified process flow diagram of the alkylation technology with H_2SO_4, licensed by STRACTO Engineering.

Olefins feed stream goes to a coalescer to remove water; after the mixture with isobutane recycling, in the sequence, the mixture is sent to the reactor. The mixture of hydrocarbons and acid follows a settler where the phase separations occur; the organic phase is sent back to the reactor, and a control valve promotes the necessary pressure reduction to vaporize the lighter hydrocarbons and remove heat from the reactor, controlling the equipment temperature, which rises due to the exothermic characteristics of alkylation reactions.

The hydrocarbons blend is sent to a flash drum where the lighter phase is directed to a compressor to condense in an accumulator vessel and the propane is recovered in the depropanizer tower while the heavier hydrocarbons (essentially isobutane) are recycled

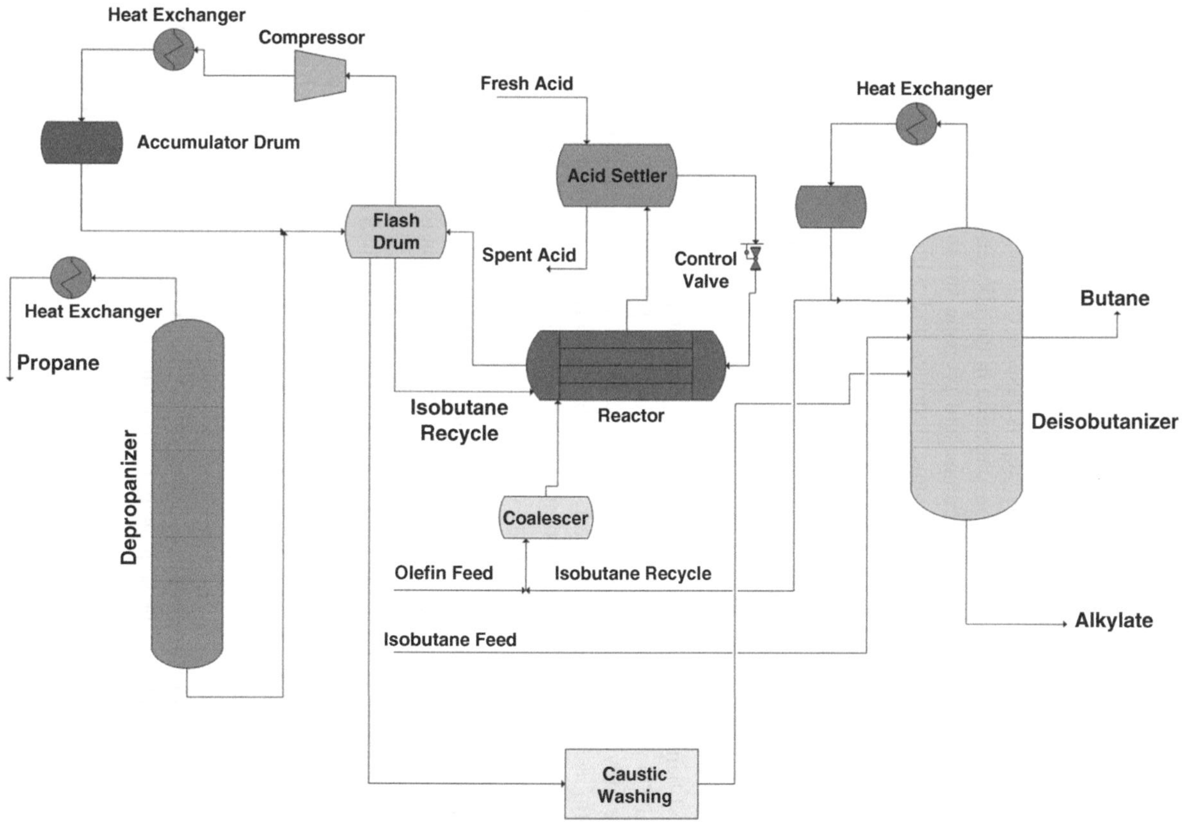

FIGURE 2.17 Basic process flow diagram of the alkylation process catalyzed by H_2SO_4, developed by STRATCO Engineering Company.

to the reactor. The stream containing the alkylate is directed to a caustic treatment and posteriorly to a deisobutanizer column where the alkylate is removed in the bottom.

As aforementioned, the need for catalyst replacement is higher in the process with sulfuric acid; however, the HF process needs a higher isobutane/olefins ratio, which means a greater separation system. Over the last decades, the refiners have opted for HF alkylation technologies due to the higher simplicity of this process and the lower need for catalyst replacement, which leaves lower operational costs.

However, regulatory pressures have led some refiners to convert their HF alkylation units to operate with H_2SO_4. Due to the high volatility and higher risks presented by the hydrofluoric acid, some licensors developed technologies to convert HF units to operate with sulfuric acids, such as the ALKYSAFE™ technology, licensed by STRATCO Engineering Company, and the ReVap™ process, developed by the companies EXXON MOBIL and CONOCO-PHILLIPS, which use additives to reduce the HF volatility, making the unit operation safer.

The principal process variables of the alkylation process are the isobutane/olefins ratio, reaction temperature, acid/hydrocarbon ratio, acid purity, residence time in the settler, and operation pressure.

For alkylation processes with HF, the reaction temperature varies from 20 to 40°C, while sulfuric acid processes operate under lower temperatures, between 4 and 10°C; under higher temperatures, the sulfuric acid can suffer decomposition to SO_2–SO_3. Operating pressure is generally sufficient to keep the hydrocarbons in the liquid phase, normally lower than 5 bar. The residence time in the settler is important because if this time is quite low, undesirable reactions can occur as organic fluoride production in the case of HF, which raises the catalyst consumption and reduces the alkylate production.

Acid purity must be maintained as high as possible through the removal of acid-soluble oil (ASO), water, and dissolved reactants in the HF case and through fresh acid replacement in processes with sulfuric acid as the catalyst.

The main disadvantage of the alkylation processes with homogeneous catalysts (HF or H_2SO_4) is the need to handle strong acids highly concentrated, which leaves greater process safety risks and high maintenance costs, mainly related to avoiding corrosion in piping and equipment.

Aiming to eliminate these risks, some licensors have dedicated their efforts to developing heterogeneous catalysts that can replace the strong acids in the alkylate production processes. The UOP Company developed a process called ALKYLENE™, which applies a solid catalyst with continuous regeneration during the process. Among other technologies that can be cited are the LURGI EUROFUEL™ process, developed by Lurgi Engineering Company in cooperation with SUD-CHEMIE; the ALKYCLEAN™ process, developed by the companies ABB-Lummus and Akzo Nobel; the FBA™ process developed by Haldor Topsoe; and the K-SAAT™ process developed by KBR Company.

2.6 NAPHTHA ISOMERIZATION

Reformed naphtha, produced in the catalytic reforming unit, is one of the streams that contribute to raising the octane number in the final gasoline. However, due to the severe restrictions related to the carcinogenic aromatic emissions, mainly benzene, some refiners have avoided the application of this stream to formulate gasoline, directing the reformed naphtha preferably to petrochemical intermediates production in aromatics complexes.

An alternative to the reforming naphtha is the production of branched hydrocarbons (with high octane number) through processes such as catalytic alkylation and isomerization.

The isomerization process involves the conversion of normal paraffin to branched paraffin, keeping the carbon number in the reactant molecule. The reactions are carried out in the presence of hydrogen under pressure and mild temperature conditions. The presence of hydrogen aims mainly to avoid coke deposition on the catalyst.

Isomerization reactions (2.1) are slightly exothermic, that is, they are favored under lower temperatures. To balance the kinetic requirements and thermodynamic limits, high-activity catalysts are applied, which allow operation under lower temperatures.

$$\text{N-paraffin} \leftrightarrow \text{I-paraffin} \tag{2.1}$$

The catalysts applied in the isomerization processes have bifunctional characteristics containing acids and metallic sites. The most employed catalysts are platinum-impregnated chloride-alumina, zeolitic, and oxide-based catalysts.

Isomerization process produces a light stream that can contribute to raising the octane number to the final gasoline and is practically free of contaminants like nitrogen and sulfur; however, due to the high content of lighter compounds, the isomerate stream negatively affects the gasoline fugitive emissions specification (high Reid vapor pressure), that is, the isomerate needs to be blended with heavier streams to formulate gasoline that meets the current specifications.

The typical feed stream to isomerization units is normally the lighter fraction (C5–C6) of straight-run naphtha when the objective is to produce isomerate naphtha, which will be directed to the gasoline pool. Another process that has great interest to the refining industry is the isomerization of N-butane to isobutane. This product can be directed to the feed stream to catalytic alkylation units or to produce MTBE (Metil-Tert Butyl Ether). In this case, the process feed stream is the heavier fraction of LPG obtained in debutanizer columns. The isomerization catalyst is sensitive to contaminants such as nitrogen and sulfur, and then normally, the feed passes through a treating step to reduce the contaminant content (normally hydrotreatment).

Figure 2.18 shows a process flow diagram for a typical isomerization unit.

The feed stream is mixed with hydrogen and is preheated in heat exchangers; following this, it is fed to the reactor, and the reactor effluent is cooled through heat exchange with the fresh feed and then is directed to a separation vessel where a major part of hydrogen is separated from the liquid phase. Hydrogen is sent back to the process by the compressor, while the liquid phase is sent to a distillation column where the isomerate naphtha is the bottom product and the light products are removed at the top.

The main process variables of the isomerization processes are temperature, operation pressure, and space velocity. As aforementioned, the temperature is normally

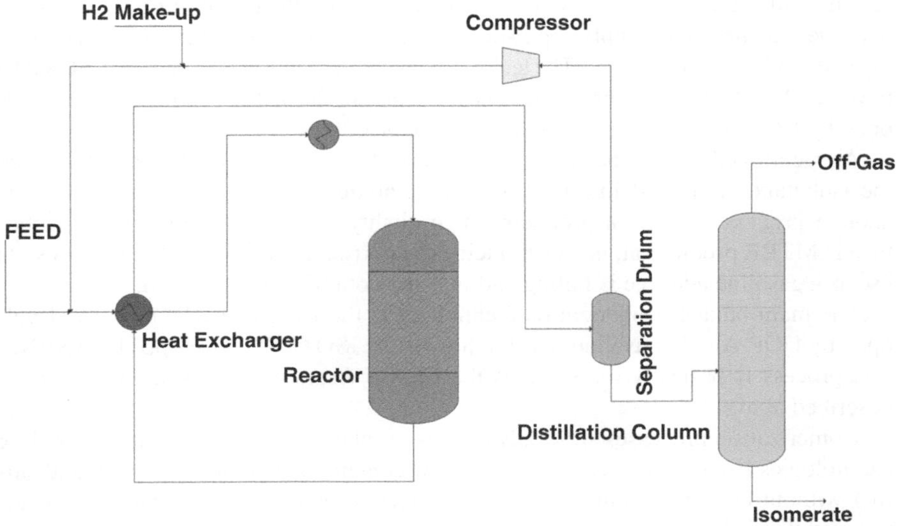

FIGURE 2.18 Typical arrangement for single reaction stage isomerization unit.

reduced and can vary between 120 and 250°C depending on the activity of the employed catalyst, due to the isomerization reactions characteristics, seeking to operate under lower temperatures.

Operating pressure is normally close to 30 bar; higher pressures conduct the higher catalyst cycle life; on the other hand, they favor hydrocracking reactions that are undesirable. The space velocity in the reactor is a design variable and expresses a relation between residence time and catalyst total cost. Lower velocity results in higher catalyst mass and higher residence time, which allows it to operate under lower temperature levels.

One of the most known isomerization technologies is the PENEX™ process, developed by UOP Company, which applies a platinum-impregnated chlorinated alumina as the catalyst.

The process applies to two reactors in series, and this arrangement allows one reactor to operate while the other is under catalyst replacement or maintenance.

Another very employed isomerization technology is the PAR-ISOM™ process, also developed by UOP Company. In this case, a sulfated-zirconia catalyst is applied with the possibility of regeneration in the processing unit.

GTC Technology Company developed the ISOMALK-2™ technology that also applies platinum-impregnated sulfated-zirconia as the catalyst. The process still applies a pre-fractionating step and the recycling of poor octane index streams.

Despite the need for higher investment, the ISOMALK-2™ technology can produce isomerate naphtha with a higher octane index.

As the isomerization reactions are limited by equilibrium conversion, some technologies involve removing isoparaffin formed during the process from the recycle stream, shifting the reactional equilibrium, and raising the quality of the final product. These technologies apply sophisticated separation processes like simulated moving beds and molecular sieves. The technologies such as IPSORB™ and HEXSORB™ developed by AXENS Company and the technologies MOLEX™ e ISOSIV™ developed by UOP company are examples of these processes.

As mentioned above, the N-butane isomerization is economically attractive once the isobutane is applied like feedstock to catalytic alkylation processes, which is another process capable of producing high-quality gasoline. Another use of isobutane is MTBE production; however, facing the current environmental regulations, its use as a gasoline additive is falling and in some countries is prohibited.

The main butane isomerization technology is the BUTAMER™ process, developed by UOP. Another available technology is the ISO C4™, developed by AXENS. The process is basically the same as the C5–C6 fraction isomerization process as described above.

Isomerization processes show advantages in relation to other gasoline upgrading technologies because they can produce a low contaminant stream (nitrogen and sulfur) without aromatic compounds, and productive processes are safer when compared with catalytic reforming and alkylation technologies, respectively.

These characteristics make the isomerization processes attractive to refiners inserted in markets with high demand for gasoline and petrochemical intermediates.

As cited earlier, relatively recent studies proved the toxicity of ethers, which led the refiners to substitute these compounds. Nowadays, the gasoline octane number is normally raised through ethanol addition or raising the percentage of naphtha from processes such as catalytic reforming, isomerization, and catalytic alkylation; however, from a technological point of view, the etherification processes are still very interesting.

2.7 LIGHT PARAFFIN DEHYDROGENATION

Light paraffin is normally commercialized as LPG or gasoline and presents reduced added value when compared with light olefins.

The dehydrogenation process involves the hydrogen removed from the paraffinic molecule and, consequently, hydrogen production, according to the reaction (2.2):

$$R_2CH\text{-}CHR_2 \leftrightarrow R_2C{=}CR_2 + H_2 \tag{2.2}$$

The dehydrogenation reactions have strong endothermic characteristics, and the reaction conditions include high temperatures (close to 600°C) and mild operating pressures (close to 5 bar). The catalysts normally applied in dehydrogenation reactions are based on platinum carried on alumina (other active metals can be applied).

Figure 2.19 shows a schematic process flow diagram for a typical dehydrogenation process unit.

The main processes that can produce streams rich in light paraffin are physical separation processes, such as LPG from atmospheric distillation and units dedicated to separating gases from crude oil.

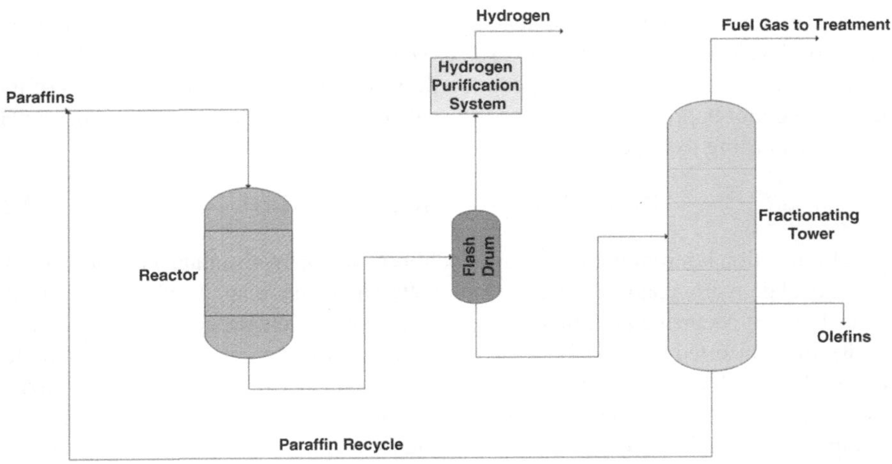

FIGURE 2.19 Process flow diagram for a typical light paraffin dehydrogenation process unit.

The feed stream is mixed with the recycle stream before entering the reactor. The products are separated in fractionating columns, and the produced hydrogen is sent to purification units (normally PSA units) and, posteriorly, sent to consumer units as hydrotreating and hydrocracking, according to the refining scheme adopted by the refiner. Light compounds are directed to the refinery or petrochemical complex fuel gas pool after adequate treatment, while the olefinic stream is directed to the petrochemical intermediates consumer market.

During the dehydrogenation process, there is a strong tendency to coke deposition on the catalyst surface, and, periodically, the regeneration of the catalytic bed is carried out through controlled combustion of the produced coke. Some process arrangements present two reactors in parallel and aim to optimize the processing unit's operational availability; in these cases, while a reactor is in production, the other is in the regeneration step.

The UOP company developed and commercialized the OLEFLEX™ which is capable of producing olefins from paraffin dehydrogenation with a continuous catalyst regeneration process. Despite the higher initial investment, this technology can minimize the unavailability period to regenerate the catalyst.

Another available technology is the CATOFIN™ process, licensed by Lummus Company, as aforementioned; in this case, two reactors in parallel are applied.

Other dehydrogenation technologies available are the STAR™ process commercialized by ThyssenKrupp-Uhde Company and the FBD™ process by SnamProgetti Company.

Due to their chemical characteristics, olefinic compounds can be employed in the production of a large number of interesting products such as polymers (polyethylene and polypropylene), propylene oxide, and oxygenated compounds production intermediates (MTBE, ETBE, etc.).

As a process of high energy consumption, there is a great variety of research in the sense of developing more active and selective catalysts that reduce the need for energetic contribution to the dehydrogenation process. One of the main variations of the dehydrogenation process is the process called oxidative dehydrogenation, which occurs according to Reaction 4.4:

$$R_2CH\text{-}CHR_2 + O_2 \leftrightarrow R_2C{=}CR_2 + H_2O \quad (2.3)$$

This reaction is strongly exothermic, and this is the main advantage in relation to the traditional dehydrogenation process. On the other hand, the main problem is the strong probability of occurring paraffin combustion against the dehydrogenation reaction.

Refiners inserted in a market with high demand for petrochemical intermediates can find in the dehydrogenation technologies a strong ally in the sense of minimizing the yield in reduced added value products and raising the profitability. In increasingly competitive scenarios, these alternative routes acquire strategic character in the perennially crude oil downstream industry.

2.8 HYDROTREATING PROCESSES

Over the years, in the face of the rising pollution levels associated with technological development and the rise in petroleum consumption, environmental legislation has become increasingly severe.

Restrictions on SOx and NOx emissions induced the necessity for higher technology development that can allow reducing the contaminant levels in the petroleum derivates mainly sulfur and nitrogen. Normally, the concentration of contaminants increases with the density of the petroleum derivate.

A lot of technologies were applied to reduce the contaminant levels in the petroleum derivates, for example, the kerosene treatment with clay, the adsorption of sulfur compounds over black carbon, and the recognized treatments by Bender and Merox. The mentioned technologies show limitations, mainly when the concentration of contaminants is high.

The hydrotreating technology (treatment with hydrogen) has been studied by many researchers in the refining industry and academic sector over the decades and, currently, it is practically impossible to meet the petroleum derivates specifications without these streams passing through the hydrotreating unit.

The hydrotreating process involves a series of chemical reactions between hydrogen and organic compounds containing contaminants (N, S, O, etc.). According to the target contaminant of the hydrotreating, the process can be called hydrodesulfurization (removing S), hydrodenitrogenation (removing N), hydrodeoxygenation (removing O), or hydrodearomatization when the main objective is to saturate aromatic compounds, among others.

The most common hydrotreating forms are hydrodesulfurization (where the objective is to remove compounds such as benzothiophene and dibenzothiophene) and hydrodenitrogenation (removing porphyrins, quinolines, etc.). These compounds, besides provoking emissions of SOx and NOx when they are burned, produce in the derivates acidity, color, and chemical instability.

The main chemical reactions associated with the hydrotreating process can be represented as below:

$$\text{R-CH} = \text{CH}_2 + \text{H}_2 \rightarrow \text{R-CH}_2\text{-CH}_3 \left(\text{Olefins saturation}\right)$$

$$\text{R-SH} + \text{H}_2 \rightarrow \text{R-H} + \text{H}_2\text{S} \left(\text{Hydrodesulfurization}\right)$$

$$\text{R-NH}_2 + \text{H}_2 \rightarrow \text{R-H} + \text{NH}_3 \left(\text{Hydrodenitrogenation}\right)$$

$$\text{R-OH} + \text{H}_2 \rightarrow \text{R-H} + \text{H}_2\text{O} \left(\text{Hydrodeoxygenation}\right)$$

where R represents a hydrocarbon.

The hydrotreating process is normally conducted in fixed bed reactors, and the most applied catalysts are cobalt (Co), nickel (Ni), molybdenum (Mo), and tungsten (W), commonly in association with them and supported in alumina (Al_2O_3). The association Co/Mo is applied in reactions that need lower reactional severity, like hydrodesulfurization, while the catalyst Ni/Mo is normally applied in reactions that need higher severity, like hydrodenitrogenation and aromatics saturation.

Hydrotreating is applied in the finishing of the final products like gasoline, diesel, or kerosene, or as an intermediate step in the refining scheme in refineries to prepare feed charges for other processes like residues fluid catalytic cracking (RFCC) or hydrocracking (HCC), where the main objective is to protect the catalyst applied in these processes.

The basic process flow is like the various hydrotreating processes (hydrodesulfurization, hydrodenitrogenation, etc.); however, the process severity, determined by variables like hydrogen partial pressure, temperature, and catalyst, varies, and the contaminant removal is affected.

The hydrotreatment process units are optimized, aiming for an equilibrium between cited operational variables, because chemical reactions are exothermic, and the decontrolled raising of the temperature can negatively affect the reactional equilibrium; besides, it's possible the sintering of the catalysts. To minimize this risk, normally the hydrotreating reactors have points between the catalyst beds where hydrogen is injected at lower temperatures (quench lines) to permit better control of the reactor temperature.

Figure 2.20 shows a typical arrangement for a hydrotreating process unit with a single separating vessel.

The configuration with a single separating vessel is normally applied in lower severity units, like hydrodesulfurization units. This arrangement is possible in this case because under reduced pressures, the difference between water and hydrocarbon properties is large, and the separation process needs reducing contact areas, so a single vessel can realize the separation process.

Higher severity units, like process units dedicated to treating unstable streams (LCO, coke gas oil, etc.) or with the objective of removing nitrogen or aromatics saturation, operate with two separating vessels like presented in Figure 2.21.

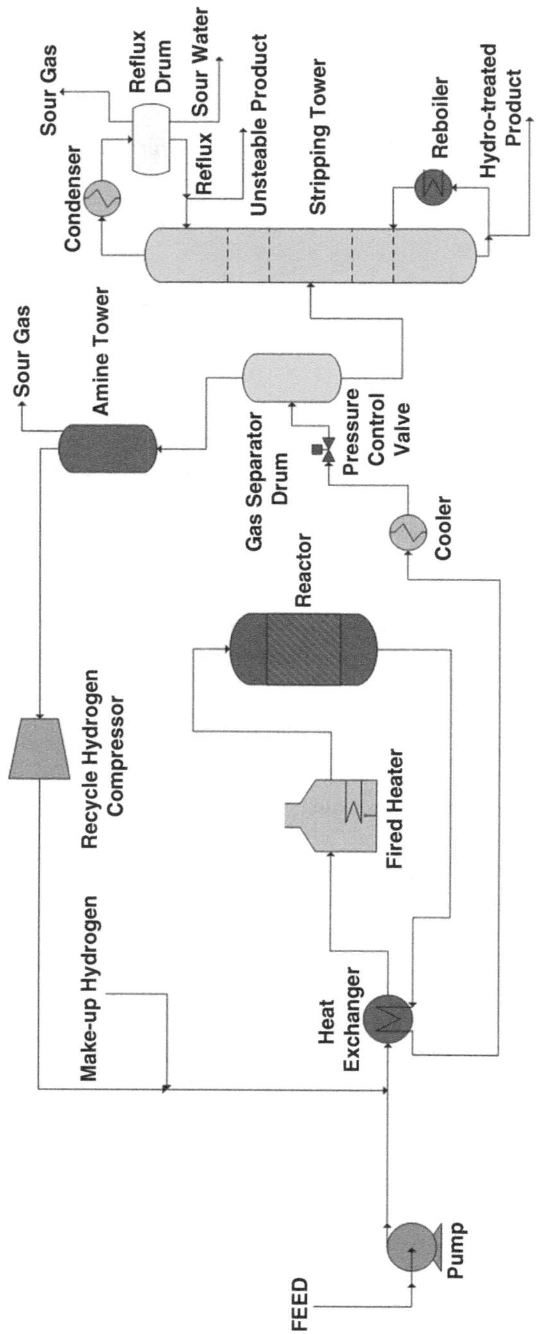

FIGURE 2.20 Basic process flow diagram for low severity hydrotreating process units.

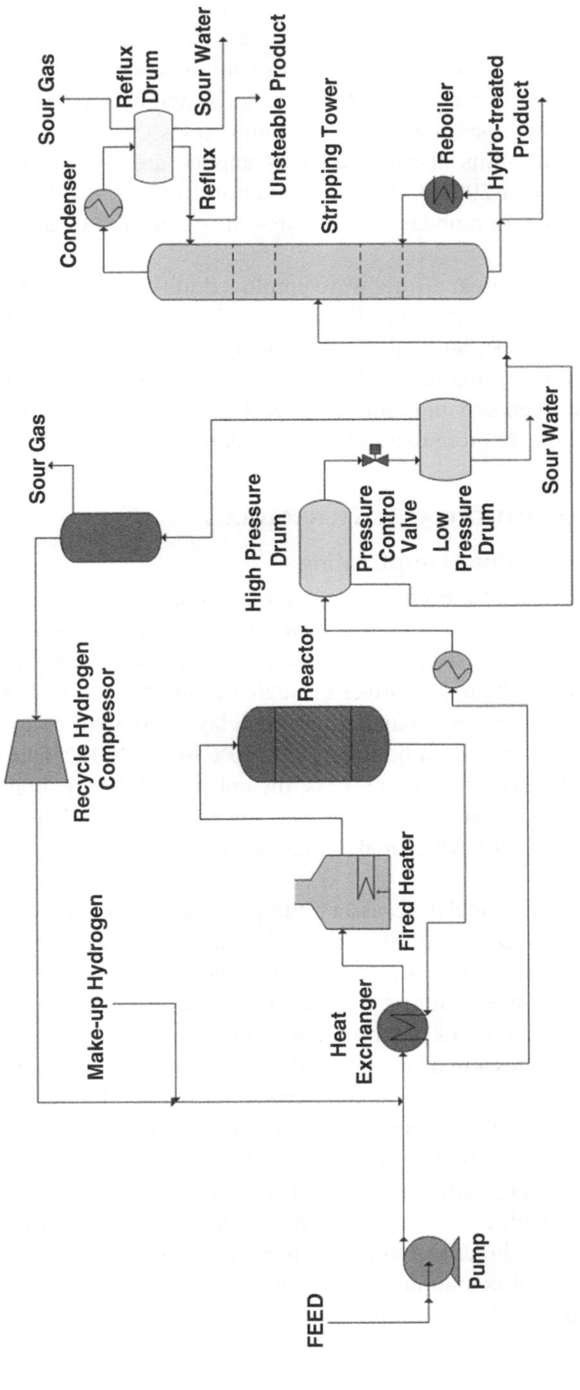

FIGURE 2.21 Basic process flow diagram for high severity hydrotreating process units.

In this case, the difference between water and hydrocarbon properties is small, and the phase separation process needs a higher interface area, so two separating vessels are applied: one under high pressure, where the separation among liquid and gaseous phases (H_2, H_2S, NH_3, and light hydrocarbons) occurs, and the other under low pressure, where the separation between aqueous and hydrocarbon phases is promoted, apart from the separation of the remaining gases.

For lower severity units, the temperatures applied are about 300 to 350°C and pressures vary between 20 and 40 bar, in addition to lower residence times. Units with high severity operate under temperatures of 350 to 400°C and pressures vary from 40 to 130 bar.

As aforementioned, great efforts were employed in the hydrotreating technology development; however, technology licensors such as Axens, UOP, Exxon Mobil, Lummus, Haldor Topsoe, and Albemarle, among others, still invest in research to improve the technology, mainly in the development of new arrangements that can minimize the hydrogen consumption (high-cost raw material) and that apply lower cost catalysts, which are more resistant to the deactivation process.

2.8.1 Naphtha Hydrotreating Technologies

2.8.1.1 Coker Naphtha Hydrotreating

Delayed coking is one of the most employed technologies dedicated to raising the recovery of high-added-value streams from crude oil. In this process, the vacuum residue from the vacuum distillation tower is converted into streams that can be used to produce middle and light derivatives through a controlled thermal cracking route.

One of the value streams produced by the delayed coking process is the coker naphtha. This stream presents a boiling range close to 40–220°C. Due to being produced through a thermal cracking process, the coker naphtha has high contaminant contents (close to 2% sulfur and 500 ppm of nitrogen) and high olefins concentration (can reach 50% in volume), which makes these streams chemically unstable (tendency to form gum).

These characteristics make necessary the previous treatment of coker naphtha before the destination to the gasoline or diesel (heavy naphtha) refinery pool. In this case, the appropriate treatment is a hydrotreating step.

Coker naphtha hydrotreating is a great challenge to the refining industry. The presence of a high quantity of olefins and contaminants leads to fouling and consequently rising in the pressure drop in the catalytic beds; furthermore, the presence of silicon-based compounds in the coker naphtha provokes the deactivation of the hydrotreating catalysts. The silicon-based compounds are added in the delayed coking units as anti-foam agents. The high olefins concentration, mainly diolefins, makes temperature control very hard in the hydrotreating reactors.

Due to these difficulties, the coker naphtha hydrotreating units are designed with three main objectives: diolefins saturation, to remove silicon-based compounds, and finally the reduction of contaminants content (hydrodesulfurization and hydrodenitrogenation). Figure 2.22 shows a process flow scheme for a typical coker naphtha hydrotreating unit.

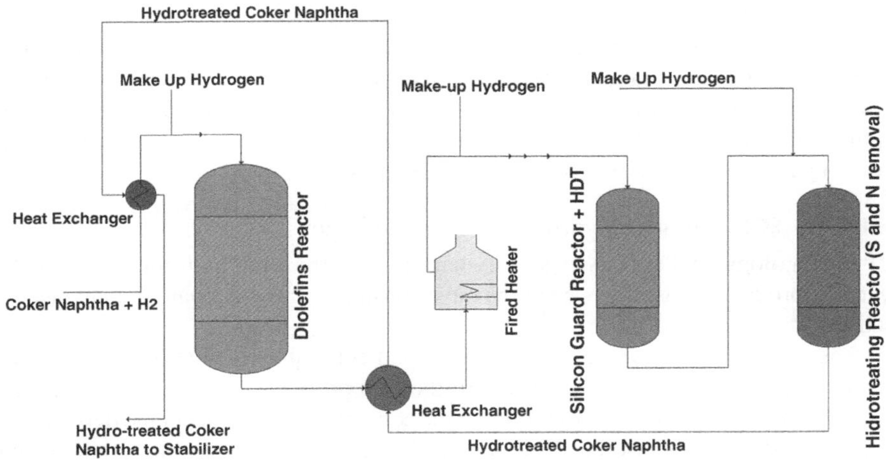

FIGURE 2.22 Typical process arrangement for coker naphtha hydrotreating unit.

The feed stream is fed to a hydrotreating reactor aimed to promote the diolefin saturation. This reactor operates under relatively low temperatures (150–240°C) due to the high heat production during diolefins saturation. In some cases, the refiners can blend the coker naphtha with straight-run naphtha to reduce the heat production; in this step, it's a good strategy to control the reactor temperature. The effluent of this reactor is fed to a second reactor that aims to remove silicon contaminants. The catalysts applied in this step have a greater porosity to avoid quick pressure drop elevation and plugging the reactor.

Next, the effluent of the second reactor is fed to another hydrotreating reactor where the hydrodesulfurization and hydrodenitrogenation reactions take place. The product from this reactor is sent to a separation system and directed to consumers, according to the refining scheme adopted by the refiner. Normally, the hydrotreated coker naphtha is fed stream to the catalytic reforming unit, but, in some cases, it can be directed directly to the gasoline pool or commercialized as a petrochemical intermediate.

In refineries that don't have coker naphtha hydrotreaters in their refining scheme, it's common to feed the coker naphtha to FCCU or for high-severity hydrotreating units (heavy coker naphtha) to compose the diesel pool of the refinery.

Due to the strategic character of this technology, some refining technology licensors, such as Axens, UOP, Haldor Topsoe, and Shell Company, developed and commercialized process technologies dedicated to treating coker naphtha. As cited earlier, the catalysts are fundamental to the success of the coker naphtha hydrotreating process; the catalyst needs to present high activity and be resistant to the high contaminants content (sulfur, nitrogen, and silicon). Some companies have dedicated their efforts to developing catalytic systems capable of meeting these requirements. As examples of these technologies, we can quote the START™ system by WR Grace Company, the UNITY™ system developed by UOP Company, and the SENTRY™ catalysts by Criterion Catalysts Company.

Coker naphtha hydrotreating is capable of raising the refiner's profitability once converting a residual stream with a difficult destination in derivatives with high added value (such as catalytic reforming feed stream) and petrochemical naphtha. Face the current scenario, with the raising of delayed coking unit installation in the refining industry, the coker naphtha hydrotreating technology acquires a fundamental role in guaranteeing competitiveness to the refiners.

2.8.1.2 FCC Naphtha Hydrotreating Technologies

The technology challenge is especially hard to treat the naphtha from FCCU, dedicated to producing gasoline, one of the most consumed crude derivatives in the world market.

The use of crude oil increasingly heavier and consequently with higher contaminant content, mainly sulfur, increased further the pressure of the regulatory agents upon the refiners in the sense of reducing the contaminant contents in the derivatives, especially diesel and gasoline. The main technology applied to reduce contaminant content in the crude oil derivatives is hydrotreating.

The conventional hydroprocessing scheme is common and widely applied in the refining industry; however, to produce high-quality gasoline, its use is limited once hydrotreating reactions fatally lead to olefins saturation, which is responsible for the high octane number in the final gasoline. Straight-run naphtha is normally directed to conventional hydrotreating because it has a reduced olefin content at the same time, which presents reduced sulfur content. In this case, the mild hydrotreating process is effective. On the other hand, naphtha from FCC tends to show high sulfur concentration and high olefins content; the processing of this stream in conventional hydrotreating units would be poorly effective.

The main chemical reactions associated with cracked naphtha hydrotreating can be represented as follows:

$$R\text{-}CH = CH_2 + H_2 \rightarrow R\text{-}CH_2\text{-}CH_3 \ (\text{Olefins saturation}) \tag{2.4}$$

$$R\text{-}SH + H_2 \rightarrow R\text{-}H + H_2S \ (\text{Hydrodesulfurization}) \tag{2.5}$$

where R is a hydrocarbon.

As aforementioned, in the case of naphtha hydrotreating to gasoline, the objective is to minimize the reaction (2.4) and maximize the yield of the reaction (2.5), aiming to keep the high octane number while the sulfur content is reduced, making the naphtha more environmentally friendly.

Over the last decades, technology process licensors have devoted their efforts to developing technologies capable of achieving these objectives. UOP Company commercializes the SelectFining™ technology that applies fixed bed reactors to promote selective hydrotreating of naphtha, leading to the unstable compounds (diolefins) saturation in the first reactor and the heavier olefins (higher sulfur content) saturation in the second reactor through the use of an adequate catalyst.

Nowadays, one of the most applied technologies to reduce the sulfur content in cracked naphtha is the PRIME G+™ process developed by the Axens Company. The process applies to the concept of which the tendency of the sulfur to concentrate in the heavier fractions of the cracked naphtha, therefore, is carried out as a feed stream fractionation before the hydrotreating step.

The feed stream is fed to a hydrotreating reactor aimed to promote the diolefins saturation, and then the stream is separated into light and heavy fractions in a distillation tower. While the light naphtha is recovered in the top, the heavy fraction is removed from the bottom of the column and sent to a selective hydrotreating section, leading to a minimum octane loss. In the sequence, the hydrotreated naphtha is separated in a stabilizer column to remove light compounds; the bottom product is mixed with the light fraction, and the final product is directed to the refinery gasoline pool.

Another technology that applies selective hydrotreating to reduce the sulfur content in the cracked naphtha is the Scanfinning™ process developed by ExxonMobil Company. In this case, fixed bed reactors without feed stream fractionation are applied. Regarding the catalysts dedicated to selective hydrogenation of FCC naphtha, we can highlight the product HyOctane™ developed by Haldor Topsoe company.

The GT-DeSulph™ process developed by GTC Engineering Company applies the extractive distillation principle associated with selective hydrotreating. The use of aromatic solvents allows the removal of thiophenic compounds from cracked naphtha.

Another interesting route adopted by the licensors is the non-selective hydrotreating, followed by an octane index recovery step; the ISAL™ technology by UOP Company employs this principle. In this case, through an adequate catalyst, the olefins saturation and posteriorly is carried out by the molecular rearrangement to recover the octane index lost in the first step.

The S-Zorb™ process developed by ConocoPhillips applies the concept of chemical adsorption of the sulfur compounds on zinc oxide (ZnO) and has a continuous regeneration system of the adsorbent through controlled burning of the adsorbed sulfur compounds.

FCC naphtha is a fundamental stream for the refiners to meet the automotive gasoline requirements (minimum octane number), and the adequate treatment of this process stream is necessary to ensure the adherence to environmental regulations at the same time in raising the profitability of the refiner once the cracked naphtha yield in the FCCUs can reach over 50%.

2.8.2 Diesel Hydrotreating Units

To comply with the new regulations, diesel production requires higher severity units. Normally, the straight-run diesel is processed with unstable streams (LCO, coke gas oil, etc.) in high-severity units, where it is possible to remove nitrogen or aromatics saturation; the unit operates with two separate vessels.

The high severity is required once, in the special case of middle distillates like diesel, the presence of contaminants like sulfide, thiophenes, and aromatic sulfur compounds like dibenzothiophenes is among the most difficult compounds to remove through hydrotreating. The hydrodesulfurization reactions are favored by higher temperatures, while the hydrodearomatization reactions are favored by higher hydrogen partial pressures. For this reason, the performance of diesel hydrotreating units requires a good balance between these variables, especially for units processing streams from FCCUs, like LCO that present high aromaticity and high sulfur content.

For higher severity units like diesel hydrotreating, the difference between water and hydrocarbons properties is small, and the phase separation process needs a higher interface area, so two separating vessels are applied: one under high pressure where the separation among liquid and gaseous phase (H_2, H_2S, NH_3, and light hydrocarbons) occurs and another under low pressure where the separation between aqueous and hydrocarbon phase is promoted, apart from the separation of the remaining gases. Units with high severity operate under temperatures of 330 to 380°C and pressures vary from 40 to 120 bar.

Due to the stricter limit of sulfur content in the diesel (< 10 ppm), Ni/Mo is applied as a catalyst to diesel hydrotreating units once this catalyst presents higher activity to desulfurization reactions.

The principal process variables considered in diesel hydrotreating units are the total pressure, hydrogen partial pressure, make-up hydrogen purity, recycle gas rate, reactor temperature, and space velocity (liquid hourly velocity, LHSV). The space velocity defines the time required to achieve a desired performance of the reactions; this parameter can be defined as presented in Equation (2.1).

$$\text{LHSV}\left(\text{h}^{-1}\right) = \text{Feed Rate}\left(\text{m}^3/\text{h}\right)/\text{Catalyst Volume}\left(\text{m}^3\right) \tag{2.6}$$

The LHSV is a key parameter to hydrotreating units, not only to the design but also to the optimization of the unit once it's possible to estimate the Start of Run (SOR) temperature and control the catalyst lifecycle based on the End of Run (EOR) temperature that is normally limited by the mechanical resistance of the material applied to reactors design. Typical diesel hydrotreating units present LSHV between 0.75 and 2.5 h^{-1}. In some designs, especially for deep hydrodesulfurization units, the reaction section is separated into two stages with the removal of H_2S and NH_3 between the reaction stages, aiming to minimize the deactivation effect of these gases over the catalyst. This can be especially attractive to hydrotreating units focused on producing ultra-low sulfur diesel (ULSD).

As aforementioned, diesel is the crude oil derivative that has had the most increasing demand in the last decades. This derivative is mainly applied as a transportation fuel by vehicles equipped with diesel cycle engines, composed of hydrocarbons between C10 and C25 with a boiling range of 150°C–380°C. The diesel ignition quality is measured through the cetane number that corresponds to a volumetric percentage of cetane (n-hexadecane) in a mixture with heptamethylnonane, which burns with the same ignition quality as the analyzed diesel. The linear paraffinic hydrocarbons are the compounds that contribute most to the diesel ignition quality, raising the cetane number, while the presence of aromatics reduces this parameter and prejudices the ignition quality. Currently, the minimum cetane number of commercial diesel is 48.

Another important parameter controlled in the diesel is the plugging point that aims to control the content of linear paraffins that tend to crystallize under low temperatures, prejudicing the fuel supply to the engine. The plugging point is determined according to the weather conditions in the region of application. In Brazil, the plugging point is controlled in the range of 0–10°C. In colder regions, the cold flow properties tend to be a significant concern to refiners, especially those processing light and paraffinic crudes like North American shale oils. In these cases, the refiners normally install dewaxing beds in the hydrotreating reactors containing catalysts based on zeolites to promote the cracking of longer paraffin.

Diesel emissions control is carried out by managing the fuel density, aiming to control the content of heavy compounds, especially polyaromatics. Currently, the density of commercial diesel is controlled in the range of 830–865 kg/m^3. To ULSD, this parameter is controlled below 850 kg/m^3. In the last decades, there have been great efforts to reduce the environmental damage produced by diesel burning. Nowadays, environmental regulations require the commercialization of low sulfur diesel with a maximum sulfur content of 10 ppm, despite the fact that in some markets, mainly in developing countries, diesel with higher sulfur content (500 ppm) is still commercialized, but this tends to change soon.

Great efforts were employed in the hydrotreating technology development; however, technology licensors like Axens, UOP, Exxon Mobil, Lummus, Haldor Topsoe, and Albemarle, among others, still invest in research to improve the technology, mainly in the development of new arrangements that can minimize the hydrogen consumption (high-cost raw material) and that apply lower cost catalysts and are more resistant to the deactivation process.

2.8.3 Bottom Barrel Hydrotreating Technologies (Residue Upgrading)

The hard scenario faced by the whole oil and gas industry requires alternative actions to ensure maximum added value to crude oil. The downstream sector faces lower demand for crude oil derivatives and puts pressure on the refining margins and competitiveness, mainly to refiners relying on refining hardware with low complexity and poor capacity of bottom barrel conversion and capable of producing lower yields of added value derivatives. In the current scenario, the players of the downstream sector can consider processing heavier crude slates, aiming to improve their refining margins once these crudes present low cost when compared with lighter crude oils or, in some cases, there is a great availability of heavy crudes.

In this scenario, process units capable of adding value to bottom barrel streams, able to improve the quality of crude oil residue streams (vacuum residue, gas oils, etc.) or convert them to higher added value products, gain strategic importance, mainly in countries that have large heavy crude oil reserves like Venezuela, Canada, and Mexico. These process units are fundamental to comply with the environmental and quality regulations, as well as to ensure the profitability and competitiveness of refiners by raising the refining margin. The necessity to add value to bottom barrel streams raised even more after January 2020 when the IMO 2020 started, which requires a deep reduction in the sulfur content of the marine fuel oil (Bunker) from 3.5% (m.m) to 0.5% (m.m).

Available technologies to process bottom barrel streams involve processes that aim to raise the H/C relation in the molecule, either through reducing the carbon quantity (processes based on carbon rejection) or through hydrogen addition.

Technologies that involve hydrogen addition encompass hydrotreating and hydrocracking processes, while technologies based on carbon rejection refer to thermal cracking processes such as visbreaking, delayed coking, and fluid coking; catalytic cracking processes such as FCC; and physical separation processes such as solvent deasphalting units.

Another fact, which raises the relevance of residue upgrading technologies, is the growing demand for petrochemicals. This fact requires a higher bottom barrel conversion capacity to convert residual streams to petrochemical intermediates.

The hydroprocessing of residual streams presents additional challenges when compared with the treating of lighter streams, mainly due to the higher contaminants content and residual carbon (RCR) related to the high concentration of resins and asphaltenes in the bottom barrel streams.

Higher metals and asphaltenes content led to a quick deactivation of the catalysts through a high coke deposition rate, catalytic matrix degradation by metals like nickel and vanadium, or even by the plugging of catalyst pores produced by the adsorption of metals and high molecular weight molecules on the catalyst surface. For this reason, according to the content of asphaltenes and metals in the feed stream, more versatile technologies are adopted to ensure an adequate operational campaign and an effective treatment.

To demonstrate the mechanism of catalyst plugging, Figure 2.23 presents a scheme of reactants and product flows involved in a heterogeneous catalytic reaction as carried out in the hydroprocessing treatments.

To carry out the hydroprocessing reactions, it's necessary to have the mass transfer of reactants to the catalyst pores, adsorption on the active sites to posterior chemical reactions, and desorption. In the case of bottom barrel stream processing, the high molecular weight and high contaminants content require a higher catalyst porosity, aiming to allow the access of these reactants to the active sites, allowing the reactions of hydrodemetallization, hydrodesulfurization, hydrodenitrogenation, etc. Furthermore,

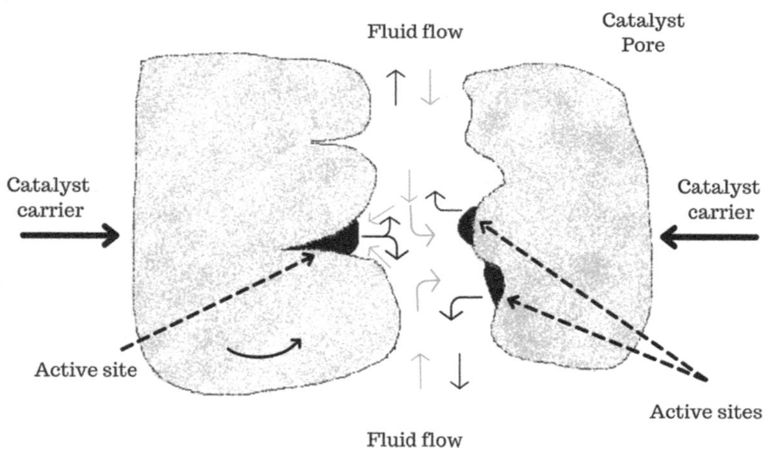

FIGURE 2.23 Reactants and products flows in a generic porous catalyst.

part of the feed stream can be in the liquid phase, creating additional difficulties to the mass transfer due to the lower diffusivity. To minimize the plugging effect in fixed bed reactors, the first beds are filled with higher porosity solids without catalytic activity and act as filters to the solids present in the feed stream, protecting the most active catalyst from the deactivation (guard beds).

The process conditions are severe in the residue hydrotreating. The feed stream characteristics lead to a strong tendency of coke deposition on the catalyst, requiring higher hydrogen partial pressures (up to 160 bar in fixed bed reactors) and higher temperatures (400–420°C).

Bottom barrel streams hydroprocessing can be applied, aiming to prepare the feed stream for other deep conversion processes like FCC and RFCC. It's also common to apply high severity hydrotreating process units to reduce the contaminants content to the processing in hydrocracking units, with the objective to protect the hydrocracking catalyst. The gas oil hydrotreating is very common in the preparation of feed streams to FCCUs, aiming to control the content of sulfur, metals, and nitrogen as well as promote the opening of aromatic rings that are refractory to the catalytic cracking reactions.

The hydrotreating process is normally conducted in fixed bed reactors, and the most applied catalysts are cobalt (Co), nickel (Ni), molybdenum (Mo), and tungsten (W), commonly in association with them and supported in alumina (Al_2O_3). The association Co/Mo is applied in reactions that need lower reaction severity, like hydrodesulfurization, while the catalyst Ni/Mo is normally applied in reactions that need higher severity, like hydrodenitrogenation and aromatics saturation. For the hydrotreating of bottom barrel streams (vacuum gas oil, delayed coking gas oil, etc.), due to the higher severity needed, is applied Nickel–Molybdenum (Ni–Mo) catalysts.

Among the bottom barrel streams hydrotreating technologies, we can quote the process Aroshift™ developed by Haldor Topsoe Company, the Unionfining™ process developed by UOP Company, the Hyvahl™ technology by Axens Company, and the RHU™ process by Shell Company.

Residue hydroprocessing can also be realized through hydrocracking process units according to the feed stream characteristics and the chosen refining configuration. Table 2.1 presents the main differences between the hydrotreating and hydrocracking processes.

Streams with higher contaminant content, especially metals, require treatment by hydrocracking. As aforementioned, in some refining schemes, hydrotreating units can be applied to prepare the feed stream for hydrocracking units, aiming to control the concentration of metals and nitrogen and protect the hydrocracking catalysts that normally have high costs. Figure 2.24 presents a typical hydrocracking process unit with two reaction stages.

The processing unit presented in Figure 2.8 relies on the intermediate separation of gases between the reaction stages. This configuration is adopted when the contaminant content (especially nitrogen) is high. In this case, the catalyst deactivation is minimized through the reduction of NH_3 and H_2S concentration in the reactors. Among the main hydrocracking process technologies available commercially, we can quote the process H-Oil™ developed by Axens Company, the EST™ process by

TABLE 2.1
Hydrotreating and Hydrocracking Processes Comparison

Hydrotreating	Hydrocracking
Contaminants removal (S, N, O, metals, etc.) and C–C bond saturation	Contaminants removal (S, N, O, metals, etc.), cracking of C–C bonds and reduction in molecular weight
Minimum cracking	High cracking rate
Low conversion (< 20%)	High conversion (> 50%)
Feed stream preparation for conversion units – FCC/RFCC, catalytic reform, hydrocracking, etc.	Production of final products – transportation fuels (diesel and kerosene) and lubricants
Ni/Co/Mo typical catalysts	Ni/W/Pt/Pd typical catalysts (dual character)

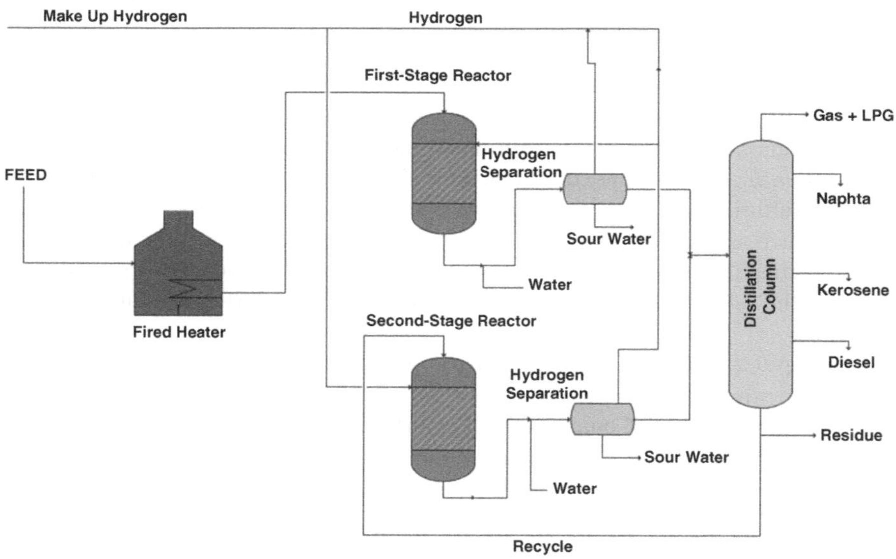

FIGURE 2.24 Typical arrangement for two-stage hydrocracking units.

ENI Company, the Uniflex™ Processes by UOP, and the LC-Fining™ technology by Chevron Company.

Catalysts applied in hydrocracking processes can be amorphous (alumina and silica-alumina) and crystalline (zeolites) and have bifunctional characteristics, once the cracking reactions (in the acid sites) and hydrogenation (in the metals sites) occur simultaneously. The active metals used in this process are normally Ni, Co, Mo, and W in combination with noble metals like Pt and Pd. The hydrocracking process is normally conducted under severe reaction conditions with temperatures that vary from 300 to 480°C and pressures between 35 and 260 bar.

It's necessary to have a synergic effect between the catalyst and the hydrogen because the cracking reactions are exothermic and the hydrogenation reactions are endothermic, so the reaction is conducted under high partial hydrogen pressures, and the temperature is controlled in the minimum necessary to convert the feed stream. Despite these characteristics, the global hydrocracking process is exothermic, and the reaction temperature control is normally made through cold hydrogen injection between the catalytic beds as well as occurring in the hydrotreating processes.

The behavior of NiMo and CoMo catalysts is strictly related to the chemical interaction between the metals and carriers (Type I and Type II catalysts) in the catalyst. The hydroprocessing reactions take place in the active sites of the catalyst, which is generally accepted to be located in the sulfur vacancies on the edges of MoS_2 crystallites; these vacancies are significantly increased when the catalyst is promoted with Co or/and Ni. The Co–Mo–S phase is similar to MoS_2 structures with promoter atoms located in the edges of a tetragonal pyramidal geometry at the edge planes of the MoS_2, while in Ni-promoted catalysts, Ni can be present in three forms after the sulfidation: Ni_3S_2 crystallites over the support, nickel atoms on the edges of MoS_2 structures, and nickel cations at octahedral or tetrahedral sites in the alumina. These different arrangements and interactions between the promoters (Ni and Co) with the MoS_2 structures and the support lead to the different behavior of CoMo and NiMo for hydrotreating reactions, with the CoMo being more selective for sulfur removal under relatively low hydrogen consumption while the NiMo catalyst is more selective for hydrogenation and hydrodenitrogenation under higher hydrogen consumption rates.

The reactivity of sulfur compounds to the hydrotreating reactions tends to be higher than that of the nitrogen compounds, as nitrogen is generally concentrated in the cracked and heavier fractions of the crude oil, and a great part of these nitrogen compounds have six or five pyridinic rings, which are unsaturated. To remove nitrogen from these heterocyclic compounds, it's necessary to hydrogenate the ring containing the nitrogen before breaking the carbon–nitrogen bond (hydrogenolysis); this is necessary due to the high energy of the carbon-nitrogen bonds in these rings. In the sulfur compounds case, the most part of the sulfur atoms are concentrated in thiophenic molecules that present relatively low energy bonds to carbon-sulfur and can directly result in sulfur removal without the necessity to saturate the heteroatom ring. For this reason, in hydroprocessing units treating heavier feeds which can concentrate refractory sulfur compounds like dimethyldibenzothiophene, the catalyst blending requires relying on NiMo bed, aiming to promote the hydrogenation function of the catalyst in order to minimize the steric hindrance of the sulfur molecules and improve the reactivity and consequently the efficiency of the hydrotreatment.

Related to the deactivation rate, this depends on the feed quality and severity of the processing unit, but it is expected that NiMo catalysts tend to have a higher deactivation rate than the CoMo catalysts once this catalyst (NiMo) is applied to treat heavier and cracked feeds, which are notable refractories to hydroprocessing reactions.

The catalyst carrier, or support, offers mechanical resistance and a high superficial area, aiming to ensure an adequate distribution of the active phase (metals), and it's responsible for controlling the acid function of the catalyst, which is desired to be low in the hydrotreating units. The support normally doesn't have catalytic activity for hydrogenation reactions, which are essentially carried out in the metal sites.

Another function of the support in hydrotreating catalysts is to ensure an adequate pore distribution, aiming to minimize the catalysts plugging due to coke or metals deposition, which can lead to a short operating lifecycle of the hydroprocessing units; this is a special concern in residue hydrotreating units.

The catalysts applied in the most severe services normally present acid and hydrogenation characteristics, especially those applied in residue hydrotreating or hydrocracking processes. Catalysts applied in hydrocracking processes can be amorphous (alumina and silica-alumina) and crystalline (zeolites) and have bifunctional characteristics more pronounced once it's desired that the cracking reactions (in the acid sites) and hydrogenation (in the metal sites) occur simultaneously. The active metals used in this process are normally Ni, Co, Mo, and W in combination with noble metals like Pt and Pd.

As described above, the acid function in hydrocracking catalysts takes place in the acidic support, which can be amorphous silica-alumina (ASA) and/or a zeolitic material, while the hydrogenation reactions are carried out in the metal sites.

Due to the severe operational conditions, the operational costs tend to be higher for the bottom barrel hydroprocessing units when compared with units dedicated to treating distillate streams (diesel, kerosene, and naphtha). The most intense hydrogenation process led to the most robust catalytic bed cooling systems (quench), higher hydrogen-replacing rates, and complex phase separation systems (multiple stages).

2.8.4 Hydrocracking Technologies

As presented in Table 2.1, for conversion rates above 50%, the hydroprocessing process can be called a hydrocracking process once a significant reduction in the molecular weight of the molecules is achieved during the process and not only the contaminants removal like in the less severe hydroprocessing units.

There are two ways to improve the quality of bottom barrel streams, the residue upgrading processes. The objective of a residue upgrading process is to raise the H/C ratio in the molecules, which can be achieved through carbon rejection or hydrogen addition.

Among the carbon rejection residue upgrading technologies, it's possible to quote the visbreaking, delayed coking, FCC, and solvent deasphalting technologies already presented in this book. The hydrogen addition route is represented by the bottom barrel hydroprocessing and hydrocracking technologies.

The hydrocracking process is normally conducted under severe reaction conditions with temperatures that vary from 300 to 480°C and pressures between 35 and 260 bar. Due to process severity, hydrocracking units can process a large variety of feed streams, which can vary from gas oils to residues that can be converted into light and medium derivates, with high value added.

Among the feed streams normally processed in hydrocracking units are vacuum gas oils, LCO, decanted oil, coke gas oils, etc. Some of these streams would be hard to process in FCCUs because of the high contaminant content and the higher carbon residue, which quickly deactivates the catalyst. In the hydrocracking process, the presence of hydrogen minimizes these effects.

According to the catalyst applied in the process and the reaction conditions, the hydrocracking can maximize the feed stream conversion in middle derivates (diesel and kerosene) and high-quality lubricant production (lower severity process).

Catalysts applied in hydrocracking processes can be amorphous (alumina and silica-alumina) and crystalline (zeolites) and have bifunctional characteristics, once the cracking reactions (in the acid sites) and hydrogenation (in the metals sites) occur simultaneously. The active metals used in this process are normally Ni, Co, Mo, and W in combination with noble metals like Pt and Pd.

It's necessary to have a synergic effect between the catalyst and the hydrogen because the cracking reactions are exothermic and the hydrogenation reactions are endothermic, so the reaction is conducted under high partial hydrogen pressures, and the temperature is controlled at the minimum necessary to convert the feed stream. Despite these characteristics, the global hydrocracking process is exothermic, and the reaction temperature control is normally made through cold hydrogen injections between the catalytic beds.

Figure 2.25 shows a typical arrangement for a hydrocracking process unit with two reaction stages, dedicated to producing medium-distilled products (diesel and kerosene).

According to the feed stream quality (contaminant content), it is necessary for hydrotreating reactors to be installed upstream of the hydrocracking reactors; these reactors act like guard beds to protect the hydrocracking catalyst.

The principal contaminant of hydrocracking catalysts is nitrogen, which can be present in two forms: ammonia and organic nitrogen.

Ammonia (NH_3), produced during the hydrotreating step, has temporary effects reducing the activity of the acid sites, mainly damaging the cracking reactions.

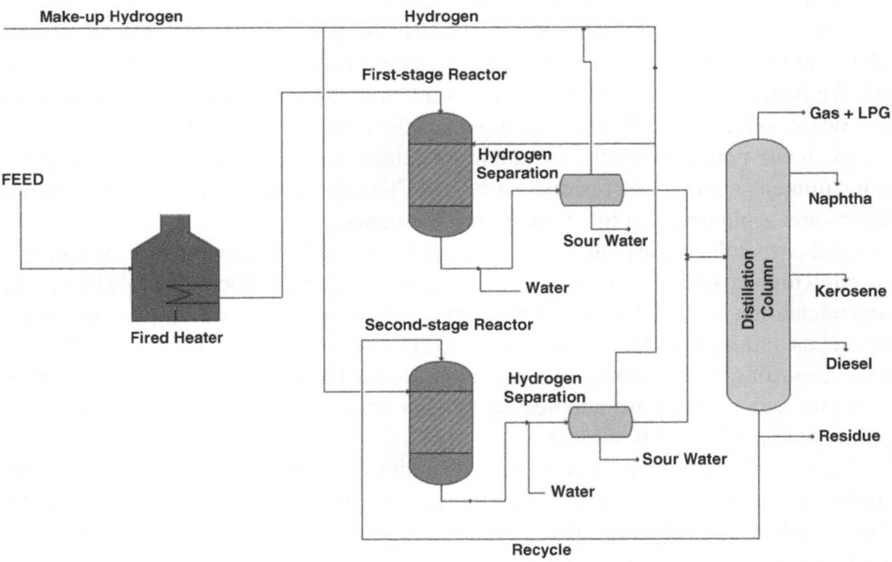

FIGURE 2.25 Basic process flow diagram for two stages hydrocracking units.

In some cases, the increase of ammonia concentration in the catalytic bed is used as an operational variable to control the hydrocracking catalyst activity. The organic nitrogen has a permanent effect, blocking the catalytic sites and leading to coke deposits on the catalyst.

As in the hydrotreating cases (HDS, HDN, etc.), the most important operational variables are temperature, hydrogen partial pressure, space velocity, and hydrogen/feed ratio.

Depending on feed stream characteristics (mainly contaminants content) and the process objective (maximize middle distillates or lubricant production), the hydrocracking units can assume different configurations.

For feed streams with low nitrogen content where the objective is to produce lubricants (partial conversion), it is possible to adopt a single stage configuration without the intermediate gas separation produced during the hydrotreating step; this configuration is presented in Figure 2.26. The main disadvantage of this configuration is the reduction of the hydrocracking catalyst activity caused by the high concentration of ammonia in the reactor, but this configuration requires lower capital investment.

Normally, for feed streams with low nitrogen content where the objective is to produce middle distillates (diesel and kerosene), the configuration with two reaction stages without intermediate gas separation is most common. This configuration is shown in Figure 2.27.

As aforementioned, the disadvantage, in this case, is the high concentration of ammonia and H_2S in the hydrocracking reactors, which reduces the catalyst activity.

The higher-cost units are the plants with double stages and intermediate gas separation. These units are employed when the feed stream has high contaminant content (mainly nitrogen), and the refinery looks for the total conversion (to produce middle distillates); this configuration is presented in Figure 2.28.

In this case, the catalytic deactivation process is minimized by the reduction in the NH_3 and H_2S concentration in the hydrocracking reactor. It's important to consider the feedstock quality to define better residue upgrading technology for the refining hardware, once the hydroprocessing of residual streams presents additional challenges when compared to the treatment of lighter streams, mainly due to the higher contaminants content and residual carbon (RCR) related to the high concentration of resins and asphaltenes in the bottom barrel streams.

Higher metals and asphaltenes content led to a quick deactivation of the catalysts through high coke deposition rates, catalytic matrix degradation by metals like nickel and vanadium, or even by the plugging of catalyst pores produced by the adsorption of metals and high molecular weight molecules on the catalyst surface. For this reason, according to the content of asphaltenes and metals in the feed stream, more versatile technologies are adopted, aiming to ensure an adequate operational campaign and an effective treatment.

As quoted earlier, the hydrocracking units demand high capital investments, mainly to operate under high hydrogen partial pressures; it's necessary to install larger hydrogen production units, which is another highly costly process. However, in the face of the crescent demand for high-quality derivates, the investment can be economically attractive.

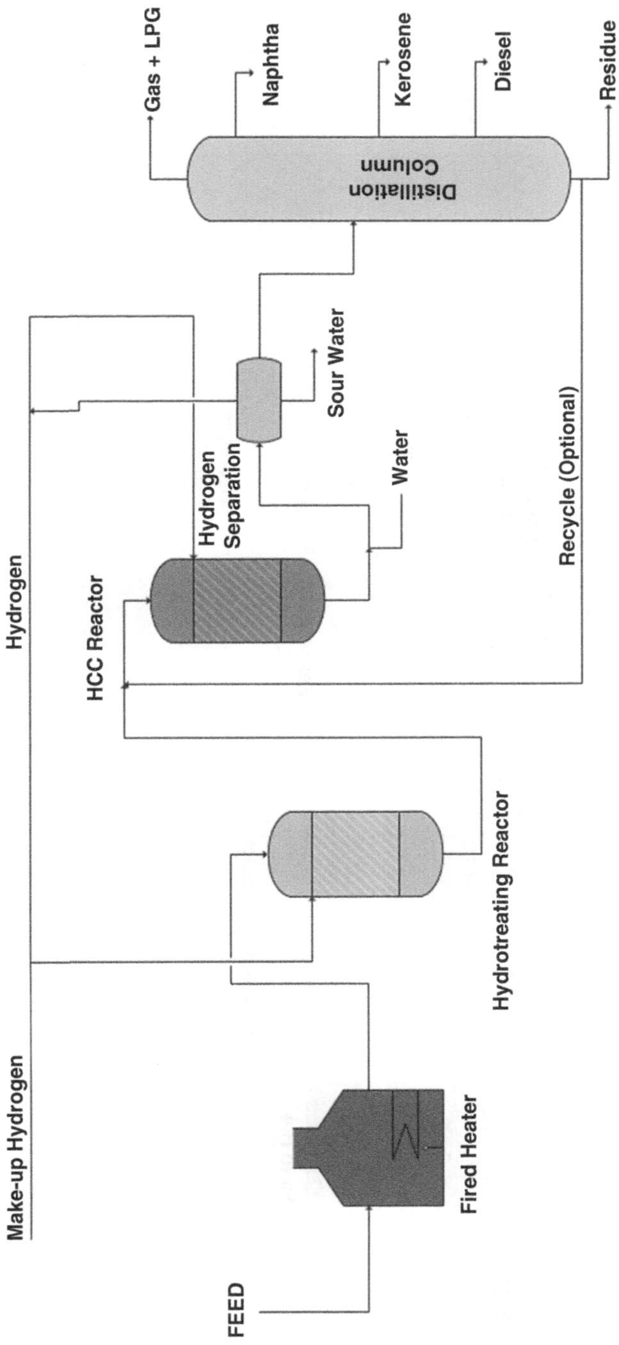

FIGURE 2.26 Typical arrangement for single-stage hydrocracking units without intermediate gas separation.

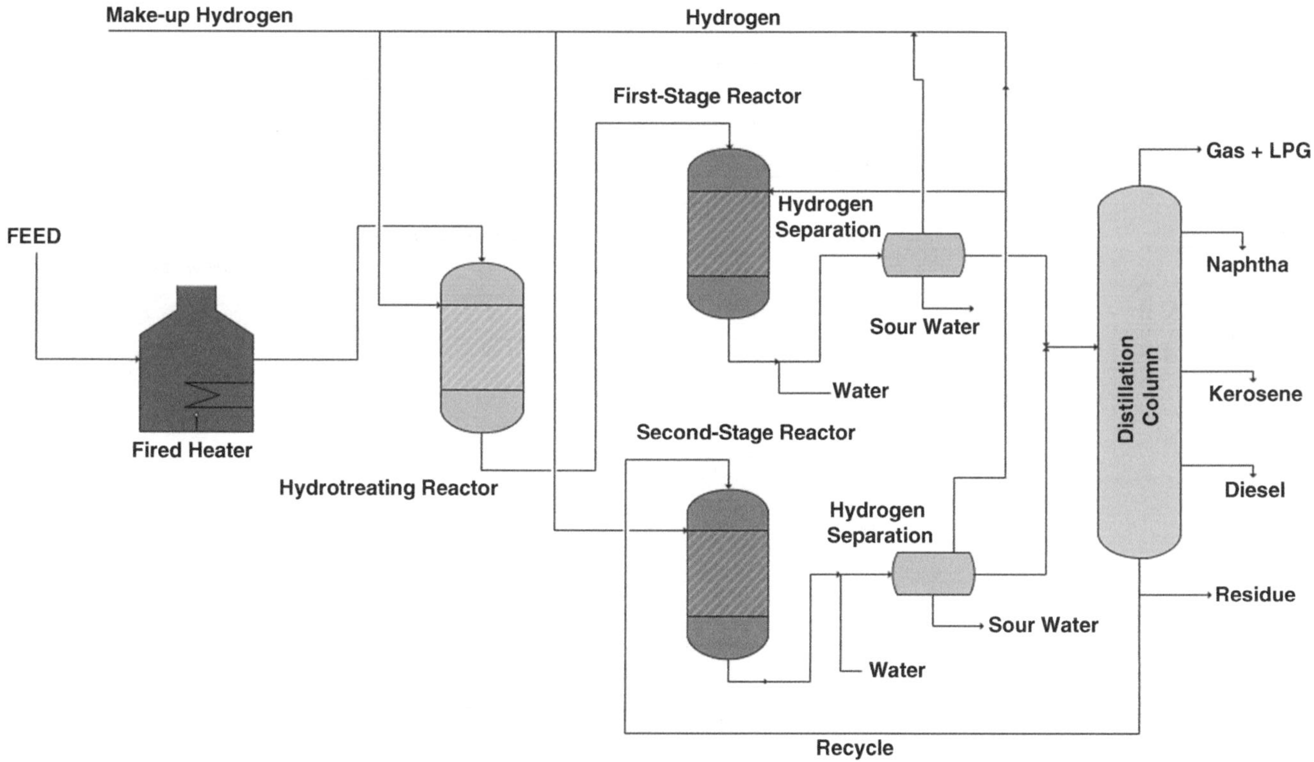

FIGURE 2.27 Typical arrangement for two-stage hydrocracking units without intermediate gas separation.

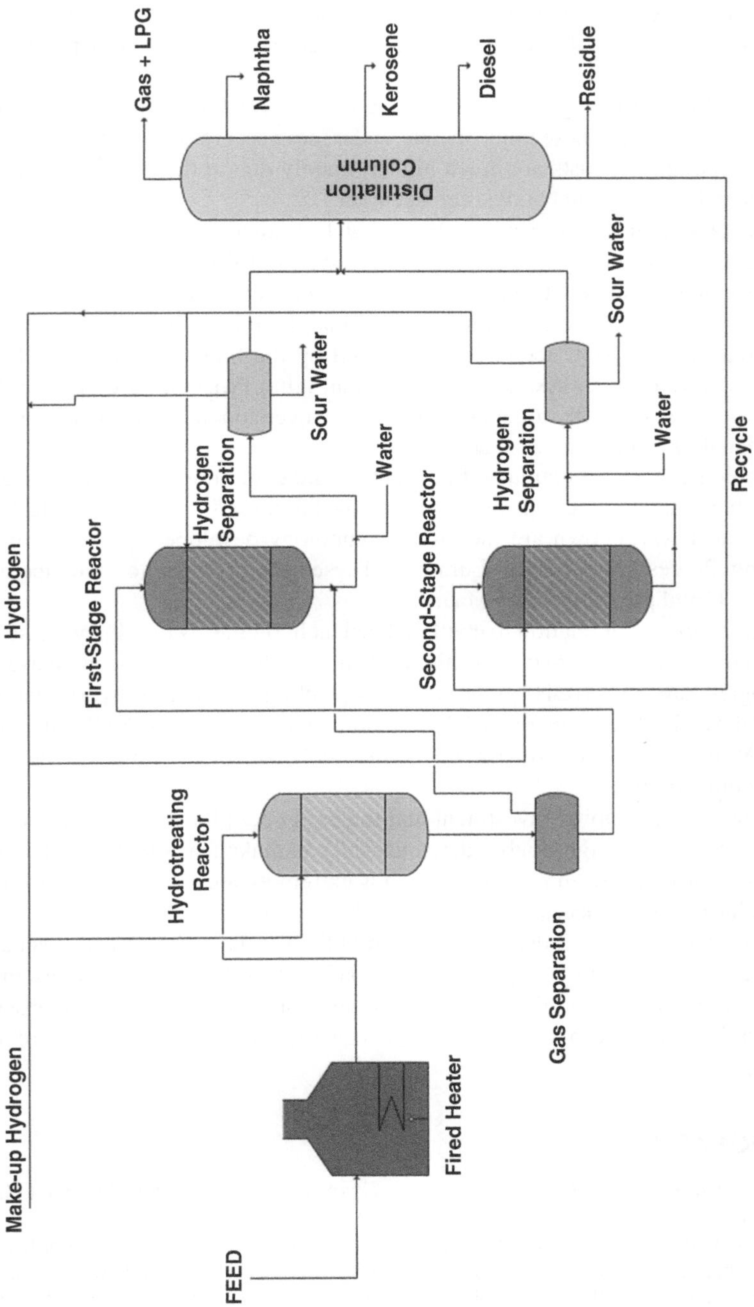

FIGURE 2.28 Typical arrangement for two-stage hydrocracking units with intermediate gas separation.

The residue hydrocracking units have severity even greater than units dedicated to treating lighter feed streams (gas oils). These units aim to improve the residue quality either by reducing the contaminant content (mainly metals) like an upstream step to other processes, such as RFCC, or by producing derivates like fuel oil with low sulfur content.

Residue hydrocracking demands even greater capital investment than gas oil hydrocrackers because these units operate under more severe conditions, and furthermore, the operational costs are much higher, mainly due to the high hydrogen consumption and the frequent catalyst replacement.

Hydrocracking technologies have been widely studied over the years, mainly by countries with large heavy oil reserves like Mexico and Venezuela. The main difference between the available technologies is the reactor characteristics.

Among the hydrocracking technologies that apply to fixed bed reactors, it can be highlighted that the RHU technology, licensed by the Shell company, the Hyvahl technology developed by Axens, and the UnionFining Process, developed by UOP. These processes normally operate with low conversion rates with temperatures higher than 400°C and pressures above 150 bar.

Technologies that use ebullated bed reactors and continuum catalyst replacement allow higher campaign periods and higher conversion rates. Among these technologies, the most well-known are the H-Oil technology developed by Axens and the LC-Fining Process by Chevron-Lummus. These reactors operate at temperatures above 450°C and pressures of 250 bar.

An improvement in relation to ebullated bed technologies is the slurry phase reactors, which can achieve conversions higher than 95%. In this case, the main available technologies are the HDH process (hydrocracking-distillation-hydrotreatment), developed by PDVSA-Intevep; the VEBA-combicracking process (VCC), developed by VEBA oil; and the EST process (Eni Slurry Technology), developed by the Italian state oil company ENI.

Despite the high capital investment and the high operational cost, hydrocracking technologies produce high-quality derivates and can make feasible the production of added-value products from residues, which is extremely attractive, mainly for countries that have difficult access to light oils with low contaminants.

In countries with a high dependency on middle distillates like Brazil (because of its dimensions and the high dependency on road transport), the high-quality middle distillate production from oils with high nitrogen content indicates that hydrocracking technology can be a good way to reduce the external dependency of these products.

BIBLIOGRAPHY

Advances in Catalysis for Plastic Conversion to Hydrocarbons. The Catalyst Group (TCGR), 2021.

Chang, R.J. Crude Oil to Chemicals – Industry Developments and Strategic Implications. Presented at *Global Refining & Petrochemicals Congress* (Houston, USA), 2018.

Fahim, M.A., Al-Sahhaf, T.A., Elkilani, A.S. *Fundamentals of Petroleum Refining*. 1st edition. Elsevier Press, Amsterdam, 2010.

Lambert, N., Ogasawara, I., Abba, I., Redhwi, H., Santner, C. *HS-FCC for Propylene: Concept to Commercial Operation*. PTQ Magazine, 2014.
Moulijn, J. A. Makkee, M. Van-Diepen, A. E. *Chemical Process Technology*. 2nd edition. John Wiley & Sons Ltd., New York, 2013.
Mukherjee, M.; Vadhri, V.; Revellon, L. *Step-Out Propane Dehydrogenation Technology for the 21st Century*. The Catalyst Review, 2021.
Myers, R.A. *Handbook of Petroleum Refining Processes*. 3rd edition. McGraw-Hill, London, 2004.
Oyekan, S.O. *Catalytic Naphtha Reforming Process*. 1st edition. CRC Press, 2019.
Robinson, P.R., Hsu, C.S. *Handbook of Petroleum Technology*. 1st edition, Springer, 2017.
Silva, M. W. *More Petrochemicals with Less Capital Spending*. PTQ Magazine, 2020.
Speight, J.G. *Heavy and Extra-Heavy Oil Upgrading Technologies*. 1st edition, Elsevier Press, Amsterdam, 2013.
Tallman, M. J., Eng, C., Sun, C., Park, D. S. *Naphtha Cracking for Light Olefins Production*. PTQ Magazine, 2010.
Vu, T.; Ritchie, J. *Naphtha Complex Optimization for Petrochemical Production*, UOP Company, 2019.
Youssef, F. Adrian, M.H. Wenzel, S. *Advanced Propane Dehydrogenation*, PTQ Magazine, 2008.
Zhou, T., Baars, F. *Catalytic Reforming Options and Practices*. PTQ Magazine, 2010.

3 The Role of Transfer and Stockpiling Operations in a Crude Oil Refinery

The main objective of a crude oil refining asset is to maximize the added value to the crude oil by converting this raw material to compounds with interesting uses for society, such as LPG, gasoline, jet fuel, diesel, asphalt, and others. To achieve this goal, a crude oil refinery applies a series of processing units in a sequence of physical and chemical processes, as presented in Figure 3.1.

A crude oil refinery is a complex organism that is completely integrated. Despite the relevance of the processing units like distillation, delayed coking, and FCC, it's impossible to carry out the refining operations without the transfer and stockpiling operations.

These operations are related to the storage and transfer of crude oil, intermediates like naphtha and gasoil, and final derivatives like gasoline and diesel. The storage of

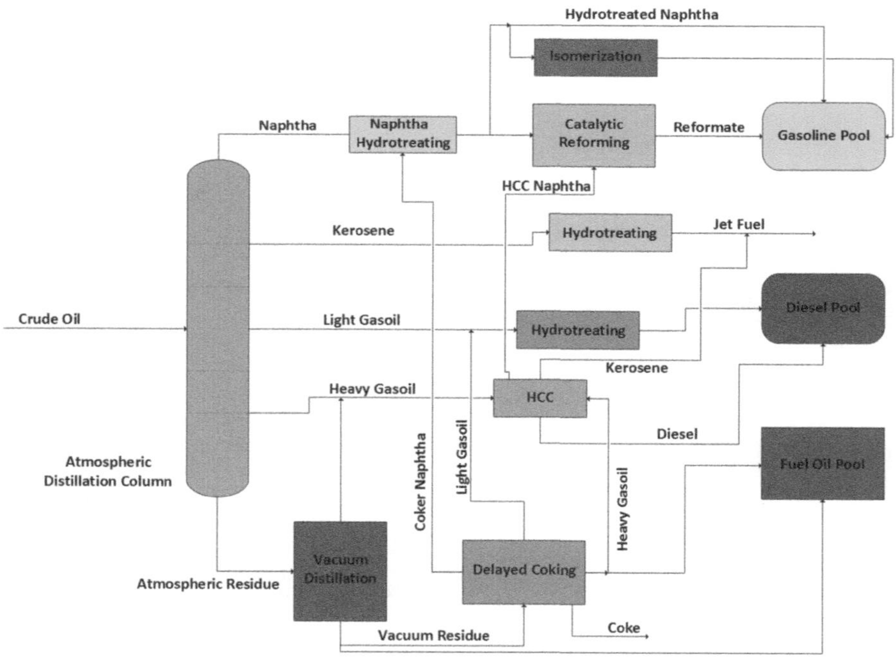

FIGURE 3.1 Typical Crude Oil Refinery.

DOI: 10.1201/9781003472216-3

these hydrocarbons is made in tanks and spheres (for liquefied gases such as LPG and propylene) adequately designed; the role of storage tanks in a crude oil refinery is to ensure capacitance to the refinery, aiming to deal with process variations with minimum production losses.

The simplified refining scheme presented in Figure 3.1 involves the interconnections between refining units and tanks which are responsible for storage intermediates and final derivatives. These interconnections are made through pipes and pumps, and these operations are the responsibility of the stockpiling operations area in a crude oil refinery. Furthermore, the stockpiling area is responsible for pumping the final derivatives to the final client, both to the local market of the refinery through local terminals or through pipelines in synergy with midstream companies.

Considering all the profitability drivers of a typical crude oil refinery, stockpiling and tank farm management have a significant impact. The effects of a poor demand by derivatives can be minimized through adequate storage capacity and adequate management, which can also be applied to minimize the necessity of additional capital investments even in scenarios of the introduction of new derivatives.

In summary, the stockpiling area in a crude oil refinery is responsible for integrating the processing units as well as starting the logistics step, aiming to ensure the market supply with the crude oil derivatives produced.

3.1 REMEMBERING FLUID MECHANICS CONCEPTS

To carry out transfer operations, pipes and tubes are applied; the main objective of these assets is to transport the fluid, and their energy is accumulated as pressure. The volumetric flow rate is defined by the amount of fluid which crosses a determined section of the pipe in a time interval and can also be defined as the multiplication of the average velocity of the fluid and the internal area of the pipe.

Considering that the fluid does not accumulate in the pipe, the volumetric flow rate is constant; under these conditions, we have a permanent flow, and the velocity increases with a reduction in the pipe diameter. Higher velocities mean higher friction losses, which lead to energy losses and consequently pressure drops.

A fluid flowing follows the energy conservation law, considering the system presented in Figure 3.2, where we have a fluid stored in a tank which will be transferred to another tank through a pipe.

Where S or Y_s is called the suction head and D or Y_d is called the discharge head. Applying an energy balance in the system presented in Figure 3.2, we can reach Equation 3.1:

$$\frac{p_1}{\rho g}+\frac{v_S^{\,2}}{2g}+y_S+H_P=\frac{p_2}{\rho g}+\frac{v_D^{\,2}}{2g}+y_D+h_L \tag{3.1}$$

where p_1 = pressure in point 1; p_2 = pressure in point 2; v_s = fluid velocity in the suction section; y_s = suction head; v_d = fluid velocity in the discharge section; y_d = discharge head; h_L = friction losses; g = acceleration of gravity; ρ = density; H_P = manometric head of the pump.

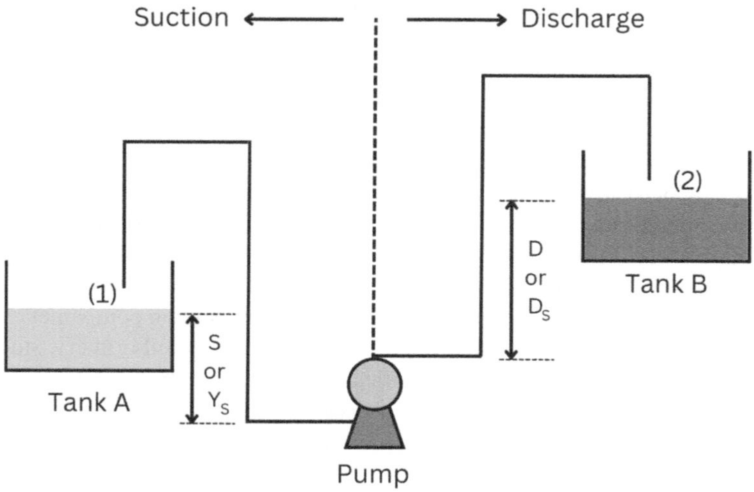

FIGURE 3.2 Typical Fluid Transfer System.

Rearranging the Equation 3.1 to isolate the manometric head of the pump:

$$H_P = \frac{p_2 - p_1}{\rho g} + \frac{v_D^2 - v_S^2}{2g} + (y_D - y_S) + h_L \tag{3.2}$$

The H_P represents the necessary energy which the pump needs to supply to the fluid, aiming to carry out the desired transfer from tank A to tank B. The energy supplied by the pump aims to overcome the friction losses and other non-idealities of the system. In Equation 3.1, the term h_L represents the total friction losses in both sections of the flow system (suction and discharge), and the calculation of this term of Equation 3.1 depends on the flow regime, which can be defined through the Reynolds number, which is defined below:

$$\text{Re} = \frac{\rho V D}{\mu} = \frac{\text{Inertial forces}}{\text{Viscous forces}} \tag{3.3}$$

where Re = Reynolds number; ρ = density of the fluid; V = average linear velocity of the fluid; D = diameter of the pipe; μ = viscosity of the fluid.

According to the literature for Reynolds numbers below 2,100, the flow regime is considered to be laminar. In a simple description, a laminar flow presents a minimum turbulence and minimum mixture of the fluid during the flow, leading to a lower friction loss. For a Reynolds number above 4,000, the flow regime is considered turbulent. For a Reynolds number between 2,100 and 4,000, the flow regime is considered under transition and can present characteristics of laminar or turbulent flow according to variables like the distance from the entrance of the tube. Back to Equation 3.1, the friction loss of the flow system can be calculated through:

$$h_L = f \frac{L}{D} \frac{V^2}{2g} \tag{3.4}$$

where h_L = friction losses; f = friction factor; L = total length of the pipe; D = pipe diameter; V = fluid velocity; g = acceleration of gravity.

The flow regime characterization is very important once determining the way of friction loss calculation and then the pump energy necessary to carry out a fluid transfer. For laminar regimes, the friction factor relies only on the Reynolds number due to the fact that the fluid flowing in the pipe creates a high-thickness boundary layer which not allow the flowing fluid to touch the pipe wall, avoiding the effect of pipe roughness in the friction factor. On the other side, for turbulent flows, the thickness of the boundary layer is very low and the roughness of the pipe begins to influence the calculation of the friction factor; the most part of the fluid transfer carried out in a crude oil refinery is conducted under turbulent flow. In this case, it's necessary to consult friction factor charts presented in the literature to determine the friction factor and then calculate the friction loss of the flow system.

3.2 PUMPS REVIEW

Pumps are the main process equipment for the transfer and stockpiling operations in a crude oil refinery, and knowledge about their operation is fundamental for the professionals in this area. Figure 3.3 presents a simplified review of the classification of the process pumps.

The volumetric or positive displacement pumps are designed to displace a defined volume of fluid in each cycle of operation. The volumetric flow rate is determined by

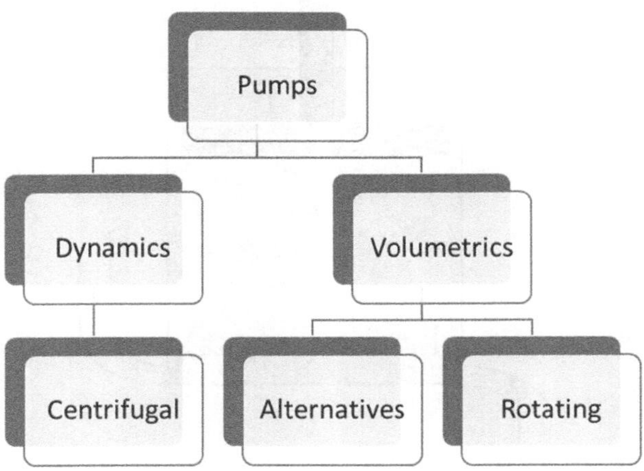

FIGURE 3.3 Pumps Classification Applied in Transfer and Stockpiling Operations of the Refining Industry.

the displacement of the moving member per cycle times the cycle rate; in other words, the flow capacity is fixed in these pumps, and the pressure or head developed by the pump depends on the flow resistance of the system in which the pump is installed and is limited only by the resistance of the components. Consequently, the discharge of a volumetric pump should never be blocked without an adequate relief system. These machines have one or more chambers, inside which a propulsive component communicates pressure energy to the liquid, causing it to flow.

The positive displacement pumps are classified as alternatives, or reciprocating pumps, and rotary pumps. Among the alternative pumps, there are the piston pumps, where the machine has a piston-cylinder arrangement, as described in Figure 3.4.

The liquid particle "a" is aspirated from point 1 (suction), follows the trajectory of the piston "P" and is discharged with pressure at point 2 (discharge).

Another very common positive displacement pump is the membrane or diaphragm pump, which is characterized by the combination of the reciprocating effect of an elastic membrane and non-return check valves to pump fluids.

In rotary pumps, the rotor causes reduced pressure at the inlet, allowing the liquid to be admitted to the pump through the effect of external pressure, and the liquid is ejected on the discharge side. As the element rotates, the liquid is trapped between the rotor components and the pump casing. Finally, after a certain rotation of the rotor, the liquid is ejected from the discharge side of the pump.

Among the rotary pumps, we can quote the lobe pumps, gear pumps, the screw, and the vane pump.

Dynamic pumps are characterized by using an impeller to transfer energy to the fluid. These pumps are normally used for moderate to high discharge rates where the

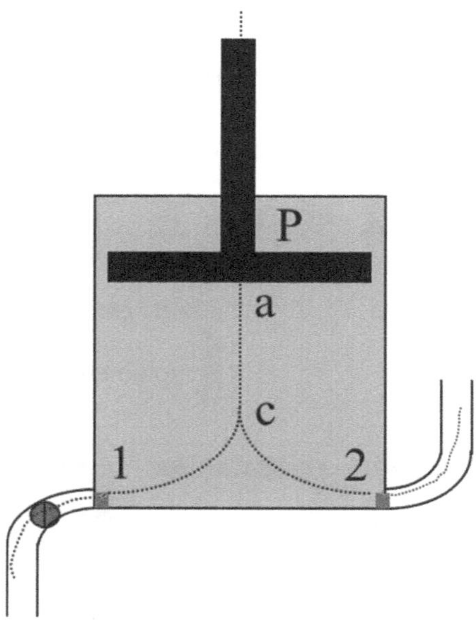

FIGURE 3.4 Typical Arrangement of a Piston Pump.

pressure differential is in a range of low to moderate. They are popularly used in a system where low-viscosity fluids are used.

The most common dynamic pumps used in the crude oil refining industry are the centrifugal pumps. In these machines, the kinetic energy is transferred to the fluid through an impeller which communicates acceleration to the fluid, converting mechanical energy into kinetic energy. The fluid enters the pump impeller along the rotating axis and is accelerated by the impeller, flowing radially through the volute chamber or the diffuser, which is connected to the piping system.

To size a pumping system, it's necessary to know the pumping system curve or the behavior of the pressure developed by the system varying the flow rate; a typical pumping system curve is presented in Figure 3.5.

This curve determines the amount of energy per weight of fluid necessary to overcome the restrictions of the pumping system for each flow rate. Considering this concept, the pumping system curve is modified by changing the characteristics of the pumped fluid, temperature, suction, and discharge levels, as well as pressure at these points, and the characteristics of the pipes and restrictions of the pumping system.

The pump curve represents the energy per weight of fluid which the pump can supply for each flow rate. A typical pump curve is presented in Figure 3.6.

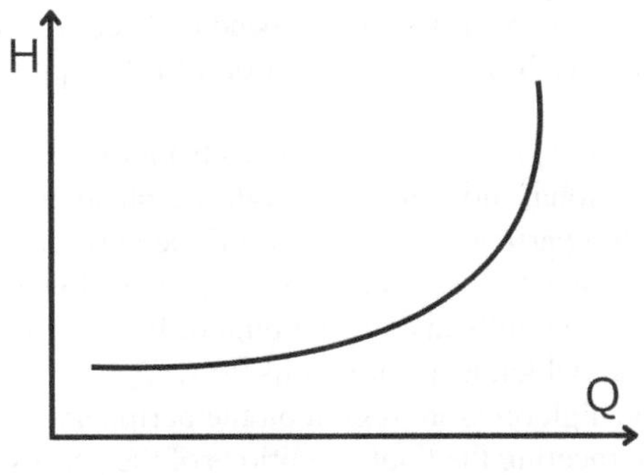

FIGURE 3.5 Typical Pumping System Curve.

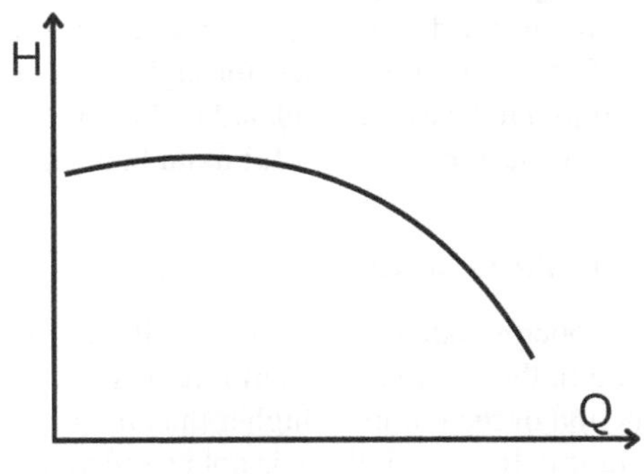

FIGURE 3.6 Typical Pump Curve.

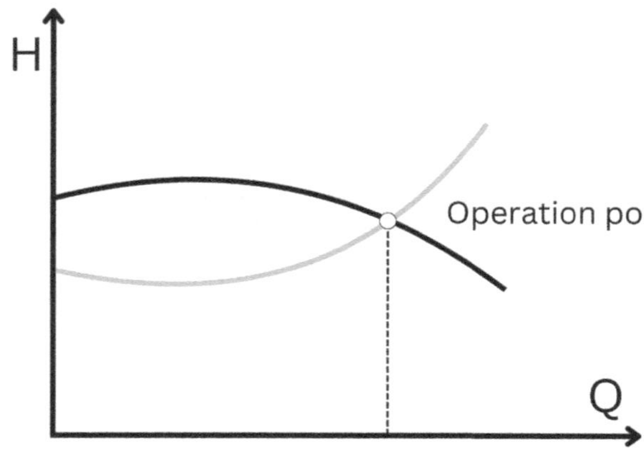

FIGURE 3.7 Concept of Operating Point for a Centrifugal Pump.

The operating point is determined by the point where the energy supplied by the pump is exactly demanded by the pumping system in order to overcome their restrictions. In other words, the operating point is the crossing point between the pumping curve system and the pump curve as presented in Figure 3.7.

To operate adequately, centrifugal pumps need to be completely filled with liquid. Its operation is practically based on the creation of a low pressure zone and a high pressure zone.

The creation of the low pressure region arises from the fact that the liquid receiving the impeller's rotational movement through the blades is subject to centrifugal force, causing the liquid particles to move toward the periphery of the impeller.

This displacement leads to the creation of a region of low pressure in the central region. This region is filled with an equal amount of liquid, thus establishing the first condition for operation, which is a continuous flow.

The creation of the high pressure region on the periphery is responsible for transporting the liquid and meeting the final conditions of the process or pumping system, since the liquid, under the action of centrifugal force, will encounter a progressive increase in the flow area that will cause a drop in speed and an increase in pressure. The impeller supplies energy to the liquid, with part of the kinetic energy then being transformed into pressure energy due to the progressive increase in the casing area in the diffusion region after the liquid passes through the volute. Considering the increasing amounts of liquid to be accommodated in the direction of flow through the volute, the volute profile is designed aiming for axial balance.

3.2.1 The Cavitation Phenomenon

As described above, to operate adequately, the centrifugal pumps need to be completely filled with liquid in the suction. Furthermore, it's necessary to have the absolute pressure in the suction of the pump be higher than the vapor pressure of the fluid at the pumping temperature. If this condition is not attended to, a portion of the liquid will vaporize, leading to the cavitation phenomenon. The bubbles produced by the liquid vaporization will be conducted to the high pressure zone of the pump and will collapse causing noise, low performance, and severe damage to the pump material.

To avoid the cavitation phenomenon, the design of centrifugal pumps needs to offer an adequate energy level in the suction section of the pump, aiming to ensure that the vaporization and consequently the cavitation will not occur. A fundamental design parameter for centrifugal pumps is the Net Positive Suction Head (NPSH), which represents the difference between the absolute energy in the suction section of the pump and the vapor pressure of the liquid at the pumping temperature. The available NPSH is defined by Equation 3.5.

$$\text{NPSH} = H_s - h_{ls} + \frac{P - P_v}{\gamma} \tag{3.5}$$

where H_s is suction level; h_{ls} is pressure drop in the suction section of the pump; P is absolute pressure in the pump suction; P_v is vapor pressure of the fluid at pumping temperature; γ is the specific weight of the fluid at pumping temperature.

Equation 3.5 presents the result of the available NPSH that the pumping system can offer to a centrifugal pump. Another key parameter which needs to be known for an adequate design of a centrifugal pump is the required NPSH. This parameter depends on the characteristics of the pump and is normally informed by the fabricant of the equipment. To avoid the cavitation phenomenon, the available NPSH should be higher than the required NPSH of the pump. This condition also determines the maximum operating flow rate of a centrifugal pump. In other words, the maximum operating flow rate is the flow rate where the required NPSH is equal to the available NPSH.

3.2.2 Association of Centrifugal Pumps

According to the service which needs to be made, it's possible to be necessary to associate two or more centrifugal pumps. In a crude oil refinery, this is not an uncommon scenario, especially for the transfer of hydrocarbons over long distances in pipelines.

There are basically two ways to associate centrifugal pumps: in series and parallel. The association in series is characterized by the discharge of the first pump and is directed to the suction of the subsequent pump as presented in Figure 3.8. The curve of the associated system is shown in Figure 3.9.

As a result of the pump's association, it's possible to operate the system with higher pressure while keeping the flow rate. Parallel association is presented in Figure 3.10.

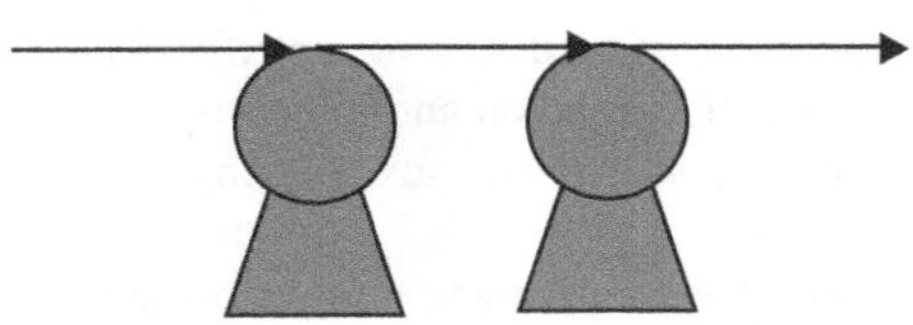

FIGURE 3.8 Association of Two Centrifugal Pumps in Series.

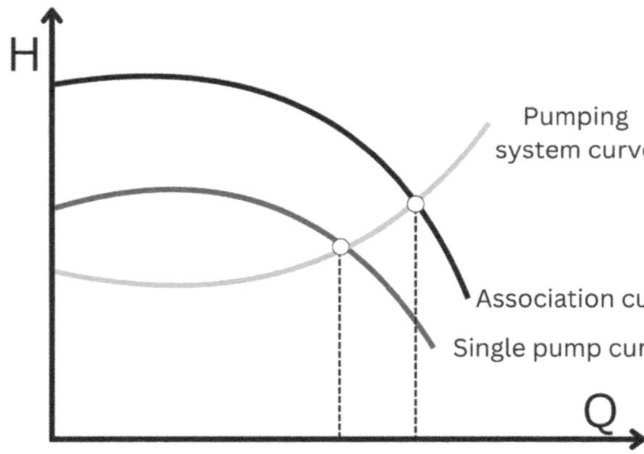

FIGURE 3.9 Association of Centrifugal Pumps in Series.

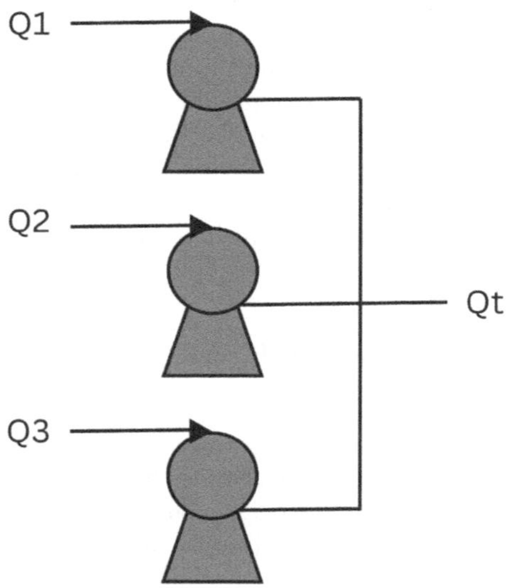

FIGURE 3.10 Association of Centrifugal Pumps in Parallel.

In this case, the flow rate is maximized while the pressure is kept constant, as presented in Figure 3.11.

It's important to avoid associating different centrifugal pumps in parallel once one of the machines can impose their head on another pump, leading to this pump operating under shut-off conditions, which can cause heating and damage to the equipment and process.

It's important to note that centrifugal pumps present the lowest start torque demanded by the driver; this is especially important for electric drivers. For this reason, centrifugal pumps with electric drivers normally start up with the discharge

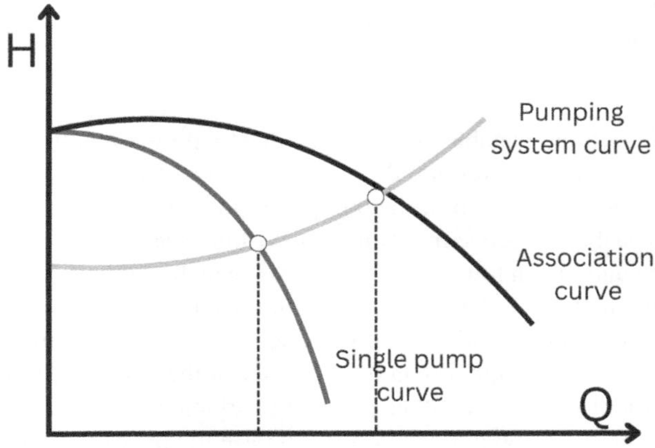

FIGURE 3.11 Association of Centrifugal Pumps in Parallel.

totally blocked or in the "shut-off" condition. In general, these key points need to be checked during the start-up of a centrifugal pump:

1. Block all drains of the pump;
2. Open the suction valve;
3. Block the discharge valve;
4. Ensure that the pump is primed;
5. Ensure that the bearing cooling system and stuffing box are aligned;
6. Ensure that the lubricating system is aligned and under oil flow;
7. Open the recirculation valve, if present;
8. Start the engine;
9. Open the discharge valve slowly after the pump reaches the nominal rotation;
10. Close the recirculation valve, if present;
11. Check if the operating system presents any leaks;
12. Check bearing temperature;
13. Monitor the behavior of the pump regarding vibration, noise, cavitation, etc.

3.2.3 Operating Issues of Centrifugal Pumps

Centrifugal pumps are the main equipment applied in transfer and stockpiling operations in a typical crude oil refinery and are relatively normal to face with operating issues of this equipment. Below, we present a short review and troubleshooting of these operating issues.

The most common issues related to centrifugal pump operation are cavitation and operation outside of the ideal operating range. The signs of cavitation are presented below:

- Noise
- Vibration

- Erosion
- Corrosion
- Suction loss.

The operating actions capable of avoiding or stopping cavitation occurrence are related to minimizing the pressure drop in the suction system, aiming to ensure that the fluid will reach the suction of the pump with maximum energy. It's necessary to check if it's possible to reduce the temperature of the suction aiming to ensure lower vapor pressure and then a lower required NPSH, which will help to avoid the cavitation phenomenon. Another key point is to minimize the pressure drop in the suction pipe by checking if all valves are totally opened. For this reason, the diameter of the suction pipe is normally higher than the diameter of the discharge pipe.

Reducing the flow rate is another operating variable which can help to avoid and stop the occurrence of cavitation as well as to raise the pressure of the suction reservoir, leading to the improvement of the available NPSH.

The signs of operation outside of the ideal operating range are presented below:

- Internal recirculation of the pump;
- Heating and energy loss;
- Accelerated wear and noise;
- Vibration

There are some operating variables capable of changing the operating point of a centrifugal pump to avoid the issues presented above. Usually, the rotation and external diameter of the impeller are varied to reach another intersection between the H x Q curve of the pump and the H x Q curve of the system, which remain unchanged, leading to another operating point to the pumping system.

3.2.4 Applying the Knowledge

As mentioned in this chapter, the stockpiling area is responsible for integrating the operations in a crude oil refinery between the storage tanks, processing units, and end consumers. For this reason, it's impossible to think of effective, profitable, modern, reliable, and safe refining assets without a well-managed stockpiling area.

In 2019, a fuel storage terminal was operating in steady state conditions. During a transfer of butane to a naphtha tank aiming to compose final gasoline, a failure in a butane transfer pump led to a hydrocarbon contention loss during a butane-enriched naphtha transfer operation to a storage tank where the seal of the transfer pump failed, the hydrocarbons were flamed, and the fire and explosion destroyed 09 storage tanks of the terminal.

The investigation conducted by the Chemical Safety Board (CSB) concluded that among the root causes of the accident was the absence of a mechanical integrity plan for the pump and the design of the storage tank farm, which was built in disagreement with safety standards like NFPA 30, contributing then to the escalation of the event.

As presented in this chapter, the operation of pumps outside of the adequate operating point can lead to vibration, which, in extreme cases, can cause severe damage to the equipment, as in this accident.

BIBLIOGRAPHY

Final Investigation Report - Storage Tank Fire at Intercontinental Terminals Company, LLC (ITC) Terminal - U.S. Chemical Safety and Hazard Investigation Board (CSB), 2023.

Karassik, I.J. and Carter, R. *Centrifugal Pumps*, McGraw Hill Book Company, 1960.

Knap, R.T., Daily, J.W. and Hammit, F.G. *Cavitation*, McGraw Hill Book Company, 1970.

Lieberman, N.P.; Lieberman, E.T. *A Working Guide to Process Equipment*, McGraw Hill Book Company, 4th edition, 2014.

McCabe, W.L.; Smith, J.C. *Unit Operations of Chemical Engineering*, McGraw Hill Book Company, 1956.

4 Types and Classification of Storage Tanks

Despite operating under continuous operations, a crude oil refinery suffers some variations and perturbations in the processes which demand time for adjustments. To deal with these variations, the crude oil refineries need to rely on a well-designed tank farm in order to ensure adequate turnaround time for the process.

This function is accomplished through the storage of crude oil, intermediate derivatives, which are normally fed to the refining processing units, and the final derivatives, such as LPG, gasoline, diesel, and jet fuel. One of the main characteristics of storage tanks is the relatively high capital investment. Figure 4.1 presents the main components of a typical hydrocarbon storage tank.

In a crude oil refinery, there are a series of tanks with different functions, and they are classified according to different criteria like operating pressure, bottom type, roof type, and shell type. Figure 4.2 presents an overview of the storage tank classification.

Regarding the roof type, the tanks installed in a crude oil refinery have fixed or floating roofs. The fixed roof is applied to tanks designed to store hydrocarbons or blend hydrocarbons with relatively low volatility or low vapor pressure, such as jet fuel, diesel, fuel, and oil. In this case, the roof is constantly attached to the tank shell and can operate under atmospheric when designed following the API 650 standard or

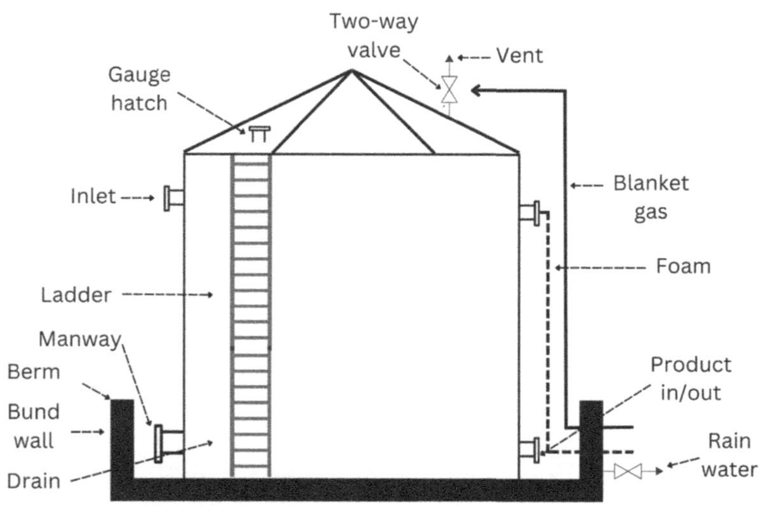

FIGURE 4.1 Main components of a storage tank.

 DOI: 10.1201/9781003472216-4

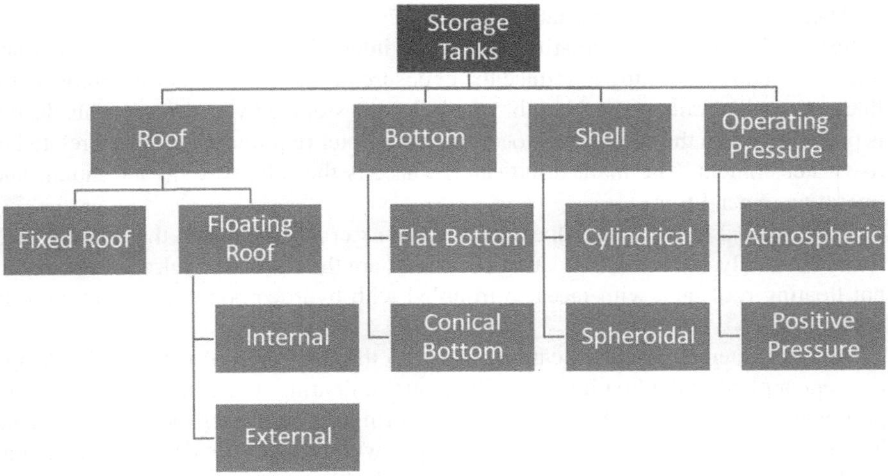

FIGURE 4.2 Classification of storage tanks.

under positive low pressure (15 psi between -325 and 250 oF) when designed following the API 620 standard.

4.1 FIXED ROOF TANKS

Fixed roof tanks are widely employed in the crude oil refining industry due to their relatively low cost of installation and maintenance. Despite these advantages, this equipment normally presents a high rate of evaporation losses and relatively high operating risks related to fire and due to the possibility of inflammable vapor cloud formation in the space between the liquid surface and the roof, this is a significant concern in scenarios of atmospheric discharges. To minimize the safety risks, the fixed roof tanks are equipped with accessories like breathing valves in order to absorb small pressure variations caused by changes in atmospheric temperature and pressure. A very important design criterion applied to fixed roof tanks is the weak welding roof mechanism, which is designed to act like a mechanical fuse in case of tank overpressure, avoiding catastrophic explosions where the tank shell is ripped off the base, causing a worsening in the emergency scenario.

4.2 FLOATING ROOF TANKS

The floating roof tanks are applied to store high-volatility hydrocarbons or a mixture of them, such as gasoline and naphtha. The use of fixed roofs here would significantly raise the risk of an explosive atmosphere inside the tank due to the evaporation of the light fraction of the liquid in the space between the liquid surface and the roof. The floating roof avoids this risk, keeping a constant pressure over the liquid surface. Another advantage in comparison with the fixed roof tanks is the lower evaporation losses. Despite this advantage, these tanks present higher capital and operating spending in comparison with the fixed roof tanks.

There are basically two variations of the floating roof tanks applied in the refining industry. The first type of floating roof is the internal floating roof, which operates under a fixed roof, aiming to reduce the exposure of the internal roof to atmospheric fluctuations like rain, preserving the quality of the storage hydrocarbon. This design is preferred when the storage hydrocarbon has stricter requirements mainly related to the water content. The main disadvantage here is the relatively higher capital and operating expenditure.

The external floating roof does not have an external roof above the floating platform. Normally, the internal floating roof tanks are the result of a retrofit of an external floating roof tank with faces so troubled with hydrocarbon contamination with water.

Another interesting classification related to the floating roof tanks is the design concept applied to the floating roof. The pontoon floating roof consists of a series of pontoons which float on the liquid surface, forming a roof over the stored product. The pontoon floating roof can be applied to a wide range of hydrocarbons such as crude oil, naphtha, and gasoline. The pontoon design presents relatively low capital cost and adequate control of the evaporative emissions.

For storage systems where evaporative emissions control is fundamental, the double-deck pontoon roof tends to be the choice. In this case, there are two decks where the first floats over the liquid and the second supports the seal. In this case, a species of air cushion is formed, which works as thermal isolation, reducing then the evaporative losses. Figure 4.3 presents a schematic of the single and double-deck pontoon floating roof.

The disadvantage of the double-deck pontoon design is the higher capital spending, and it is applied only in special situations where it is essential to control the evaporative emissions, such as benzene and methanol storage tanks.

Another type of floating roof applied in the crude oil refining industry is the buoyroof. This kind of floating roof is a variation of the pontoon design where auxiliary floats are installed over the floating deck, aiming to improve the roof buoyancy, especially in the case of water accumulation over the roof. This design is normally applied in large-diameter tanks as they are applied to store crude oil. The buoyroof floating roof is presented in Figure 4.4.

Considering the characteristics of the pontoon and buoyroof floating roofs, it's possible to conclude that the pontoon design is more adequate to deal with lighter

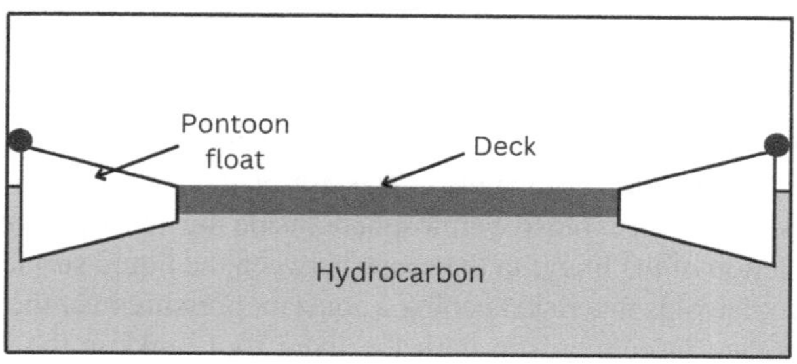

FIGURE 4.3 Schematic of the single and double-deck pontoon floating roofs.

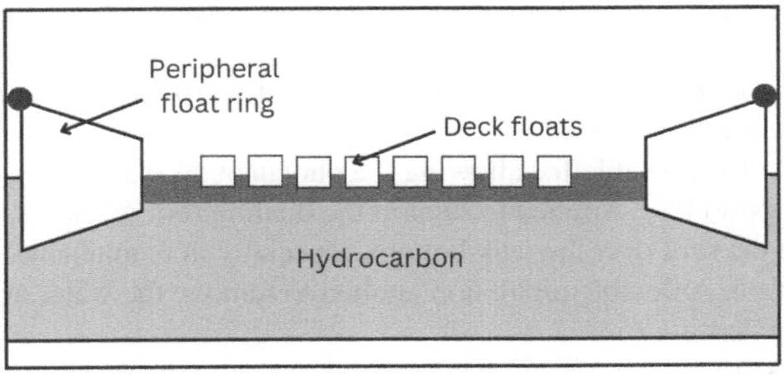

FIGURE 4.4 Schematic of a typical buoyroof floating roof.

products during the storage period, once allowing a lower roof sinking and then a lower inclination angle during the service, leading to better buoyancy characteristics. Furthermore, pontoon roofs offer higher pressure over the liquid than buoyroof roofs, which is a desired characteristic when the stored fluid presents high vapor pressure. Once it's possible to occur, the vapor accumulation under the roof can destabilize the roof; the format of the pontoon roof helps to direct this vapor accumulation to the periphery of the roof, where the temperature tends to be lower, leading to vapor condensation and then stabilizing the roof buoyancy as presented in Figure 4.5.

According to the API 650 standard, the floating roofs should have buoyancy capacity to operate with liquids with a minimum specific gravity of 0.7 and vapor pressure below 76.5 kPa.

It's fundamental to control the density and vapor pressure of the liquid which will be stored in the floating tank, aiming to avoid receiving liquids with density and vapor pressure beyond the values determined by the floating roof design. This observation is essential to avoid the roof sinking and the total loss of the storage tank.

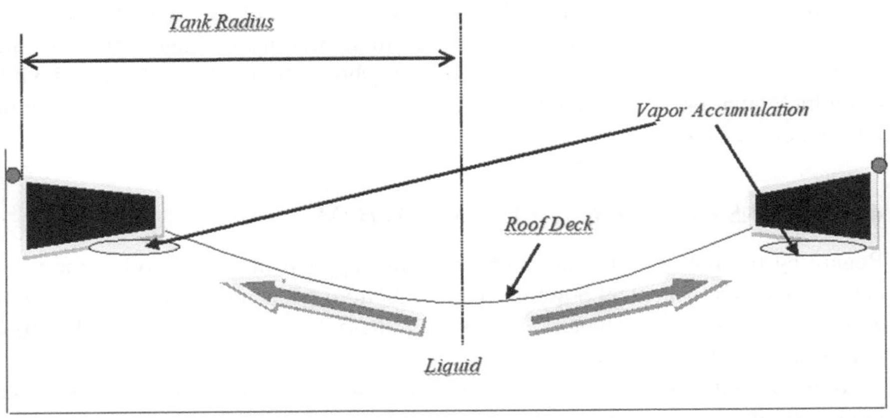

FIGURE 4.5 Behavior of pontoon floating roof with vapor accumulation upon the deck roof.

Among the main components of a typical floating roof, we can quote:

- Floating deck: The main structure of the floating roof, which is kept in contact with the stored product.
- Floaters: Responsible for allowing the buoyancy of the floating roof.
- Roof support legs: Applied to support the floating roof when it's necessary to support the roof over the tank bottom, especially in maintenance situations.
- Roof drain: A flexible tubulation applied to remove the water accumulation over the floating roof.
- Seal or sealing chamber: System applied to ensure the seal between the tank shell and the floating roof.
- Mobile ladder: System applied to allow the access of operating and maintenance personnel to the floating roof.

Especial attention needs to be paid to two components of the above list. The roof support legs can suffer severe corrosion during the operating lifecycle of the storage tanks, which is normally longer than 20 years. Before authorizing support, the roof in the bottom of the tank for maintenance, it's fundamental to request an inspection through thickness measurement of the roof support columns, aiming to ensure that these accessories can support the weight of the roof. If the roof support columns present low thickness, they can collapse, leading to loss of the roof and, in the worst case, the whole tank.

Another key accessory of floating roofs is the roof drain. This tubulation is applied to avoid rainwater accumulation over the floating roof; this accumulation can lead to the roof slopping and collapsing, leading to total loss of the tank. For this reason, it's necessary to have a maintenance routine to verify the integrity and availability of the draining system of floating roofs as well as agility to solve eventual issues. The tank should be kept out of operation until the return of the draining system.

Another key point here is the type of tubulation applied in the draining system. The old design applied articulated rigid tubulations (flexible swing joints), which present a high failure rate and can lead to contention loss of the stored hydrocarbon and risks of fire and explosion. Once it's hard to inspect the drainage tubulation during the tank operation, this is a great risk and should be minimized using flexible draining tubulation, which can be installed by opportunity during the periodic maintenance of the tanks.

4.3 TANKS CLASSIFICATION BY BOTTOM TYPE

Regarding the bottom type, normally the tanks applied in a crude oil refinery are classified as flat or conical bottoms. The flat bottom tanks are normally small, with an internal diameter of 6 meters. In this case, the bottom drainage can reach a good performance with a flat bottom.

Tanks with diameters higher than 6 meters and dedicated to storage of hydrocarbons or derivatives with stricter quality requirements, like jet fuel and aviation gasoline, demand special attention with the drainage operation and the use of conical bottom, aiming to maximize the effectiveness of the bottom drains.

Considering the stricter regulations of jet fuel, some refiners apply floating suctions in jet fuel storage tanks. The floating suction allows suction to be always near the surface, where the fuel is cleaner, and hydrocarbons do not contain water that settles in the bottom of the tank. Floating suctions are used for safe and clean fuel draw off from storage tanks where fuels need to be uncontaminated by water or solids.Floating suctions are able to siphon a product from just under the product surface, regardless of the surface level. This design avoids sucking up particles or sediment deposited at the bottom of the reservoir.

4.4 TANKS CLASSIFICATION BY SHELL TYPE

Regarding the tank shell, the main types applied in the crude oil refining industry are cylindrical and spherical shells. Cylindrical shells are the most common tanks and are designed to operate under atmospheric and positive pressure.

Positive pressure tanks are designed according to specific standards like API 620, where a positive pressure inside the equipment is expected (limited to 15 psig). An example of the application of these tanks in a crude oil refinery is the tanks dedicated to the storage of sour water from high severity hydroprocessing units. In this case, the stored fluid has a high concentration of dissolved NH_3 and H_2S, and it is necessary to keep a positive pressure over the fluid to avoid toxic emanation to the atmosphere.

The spherical shell tanks are applied to store liquefied gases such as liquefied natural gas (LNG), LPG, propane, propylene, and ethylene in the crude oil refinery or petrochemical plant. The use of spherical shells offers lower possibility of tension concentration in the equipment and minimizes the evaporation losses due to the lower heat flow, which is a consequence of the higher volume/surface ratio of the spherical shells.

4.5 TYPICAL ACCESSORIES OF STORAGE TANKS

The storage tanks demand some accessories equipment, aiming to ensure process safety and better performance to the storage and blending operations. The main accessories included in a typical hydrocarbon storage tank are:

- Venting system.
- Mixing system.
- Gauging system.
- Heating system.
- Thermal isolation.
- Draining system.
- Bund.

4.5.1 VENTING SYSTEM

During the filling operation of a storage tank, the vapor space is reduced due to the raising of the liquid level, leading to increasing pressure in the vapor space, while during emptying operations, the liquid is withdrawn from the equipment, leading to the vapor space to increase and a reduction in the pressure.

In both operations described above, it's necessary to have a system capable of allowing the internal pressure control of the tank to avoid overpressure and vacuum formation inside the tank. The vent systems are designed to avoid both scenarios, allowing air access to the tank during emptying operations and air relief during filling operations, aiming to keep the operating pressure under the design limits.

Further, in the filling and emptying operations, the overpressure and vacuum situations can occur during the heating and cooling operations of the tank, as the heating can lead to vaporization and growing pressure inside the tank, as well as the cooling process can cause a kind of contraction of the vapor space, creating a vacuum which can collapse the tank structure without an adequate and well-designed vent system. This scenario is a serious concern for storage tanks operating with heated derivatives like asphalt, which demand an operating temperature of around 150°C. If the thermal isolation of the tank is not adequately designed as well as the vent system, the tank can collapse due to sudden cooling during storms and rain.

The calculus of venting systems for low pressure and atmospheric storage tanks is based on the API 2000 standard. It's important to consider not only the filling and emptying flowrates to design the venting systems but also the thermal aspects over the tanks. The API 2000 standard defines two ways to design the venting systems:

- Simplified Approach: It considered only the filling and emptying effects to determine the air flowrate entering and leaving the tank;
- Rigorous Approach – It considers the thermal effects produced by sudden cooling of the tank sides and further the filling and emptying effects.

The simplified approach demands special attention to be applied once, which will lead to lower venting systems which cannot attend to the process demand in some scenarios. This method should be applied only if it is possible to guarantee that there are no torrential rains or sudden variations in the climate.

A key aspect of the venting systems is the use of thermal isolation in the tank side and roof, which considerably reduces the dimensions of the venting systems to be installed in the storage tanks.

4.5.2 Mixing System

In some crude oil refineries, the storage tanks are used to store and prepare the final derivatives through a blend of different streams. An example of this operation is gasoline production, which is made, in some cases, to send naphtha streams to the final storage tank. The inventory is mixed to ensure the homogeneity of the final blending and then reach the final gasoline. This process is exemplified in Figure 4.6.

Due to the different characteristics of the process streams which compose the final products, it's necessary to mix the tank inventory to ensure adequate homogeneity and compliance with the quality requirements of the final derivative.

There are two mixing systems normally applied to hydrocarbon storage tanks: the mechanical and the jet mixers.

The mechanical mixer is a machine powered by an electrical engine installed on the tank side. The mixing effect is achieved through an impeller inside the tank, as presented in Figure 4.7.

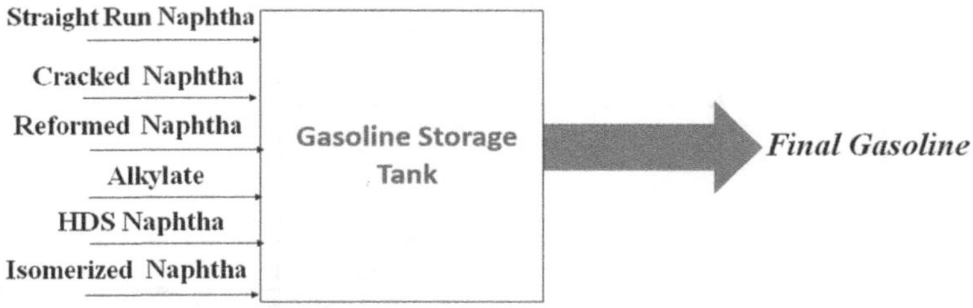

FIGURE 4.6 Gasoline production process in a storage tank.

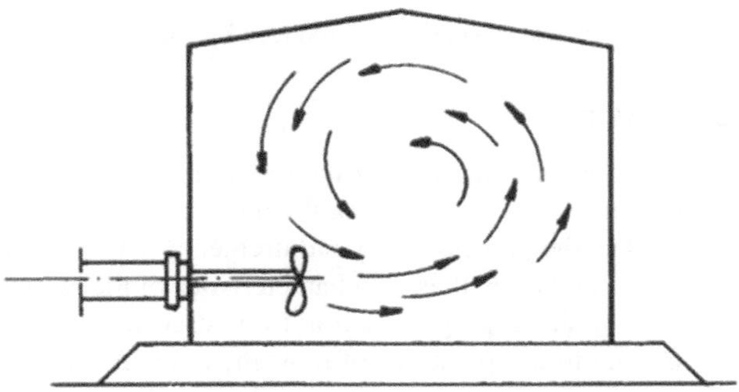

FIGURE 4.7 Schematic for the mixing effect produced by a mechanical mixer.

According to the dimensions of the storage tank and the characteristics of the stored product, the quantity of mixers can be determined to ensure adequate homogeneity of the inventory in adequate time intervals in order to meet the production plan of the refinery.

The main disadvantages of mechanical mixers are the presence of rotating parts, which can lead to vibration and leakage, which can cause serious process safety issues or the unavailability of the tank. To avoid this, these machines demand an adequate asset management program to anticipate their maintenance before failure and the consequent impact on the process and the production plan of the crude oil refinery. In some cases, this can lead to significant operating costs.

Jet mixers, or side entry mixers, use jet-flowing technologies to achieve high turbulence flow inside the storage tank and ensure the mixing effect. This process is demonstrated in Figure 4.8.

The vortex motion produced by the jet mixer allows a good mixing effect without rotating parts such as the mechanical mixers. The operating principle of the jet mixer is using the inventory of the tank in circulation using a pump to be injected back into the tank under high velocity and then ensuring the mixing effect as well as temperature homogeneity of the tank inventory.

The advantage of jet mixers over mechanical mixers is the absence of rotating parts, which minimizes maintenance dependence and leakage risks. On the other side, for dirty and high viscosity derivatives, the jet mix tubulation can be blocked, leading to the necessity to open the tank to make the cleaning and unclogging of the jet mix.

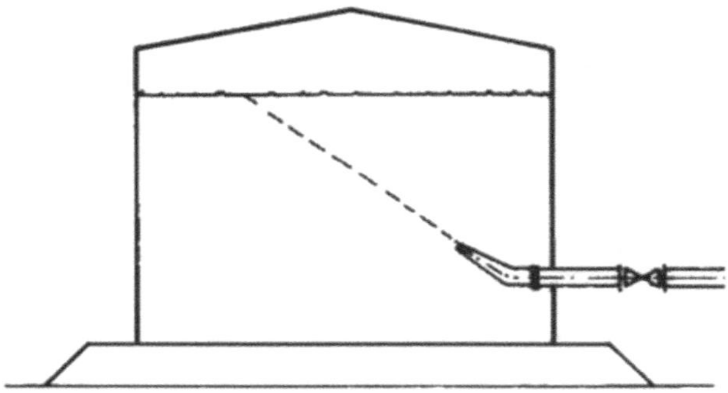

FIGURE 4.8 Schematic for the mixing effect produced by a jet mixer.

4.5.3 Gauging System

The gauging system is an essential part of the operating system of storage tanks to ensure the measurement of liquids, allowing the quantification of the produced volume and mass as well as the process safety requirements. The measurement method depends on the type of tank as well as the characteristics of the stored liquid.

The purpose of the tank gauging system is to be used in a crude oil refinery to control the inventory, monitor operations of movement and transfer of crude oil and derivatives, maintain the mass balance of the crude oil refinery, prevent overfilling, and detect leaks.

Regarding operations of movement and transfer of crude oil and derivatives, the tank gauging systems allow adequate control of the process with information such as how much liquid is in the inventory expedition tank, when the receiving tank will reach a dangerous level, and when the inventory expedition tank will reach the minimum level with a given pump rate.

The tank gauging system is also responsible for producing alerts and alarms in order to keep the operation team alert regarding the dangerous levels of the tanks. Another fundamental characteristic of the tank gauging system is allowing leak detection in the tank or the tank farm.

When a storage tank is filled and settled, it's common to set the gauging system to detect small movements of the liquid inventory. This function is normally called the "stopped tank alarm" and is fundamental to help the operation team detect undesired transfers between tanks or contention loss in the tank farm, which is a powerful safety layer to avoid and minimize process safety accidents.

Among the tank gauging technologies, the first one applied was hand gauging, which applies a specially designed measurement tape to measure the liquid level inside the tank. The hand gauging is normally used to check and calibrate the electronic gauging systems.

When automatic gauging systems were introduced in the industry, in the 1930s, the float gauge was the first system used on an industrial scale. The system consists of a float installed inside the tank connected with an external metallic tape, which has a graduated scale. Unfortunately, the accuracy of this system is low and presents poor reliability. The constant development in gauging systems led the industry to develop servo gauging systems in the 1950s. In this gauging technology, the float is replaced by a small displacer. The displacer has buoyancy but does not float on the liquid; the mechanism is suspended by a thin wire which is connected to the servo gauge on top of the tank. A weighing system in the servo gauge senses the tension in the wire; signals from the weighing mechanism control an electric motor in the servo unit and make the displacer follow the liquid level movements. The servo gauge presents higher accuracy than the float gauge, and if adequately calibrated, this system can meet the custody transfer requirements. Despite this, the system presents low reliability and relatively high failure rates.

In the 1970s, industry development reached radar gauge systems. Radar-level transmitters work with microwave pulses emitted by an antenna and reflected on the liquid surface. The wave travel time is proportional to the traveled distance, which corresponds to the liquid level.

The absence of moving parts and contact with the liquid makes it possible to use the radar gauge technology for a wide variety of liquids and makes the radar the most applied gauging technology in the last decades.

4.5.4 Heating System

This system is applied to keep the fluidity of the stored liquid, aiming to allow the transfer, in other words, to maintain the pumpability of the liquid. Normally, the heating system consists of a coil installed inside the tank where steam is fed to supply heat and then raise and keep the temperature. Nowadays, considering the growing demand for decarbonizing the crude oil refining industry, it's possible to find electrical heaters for this purpose, although with high electrical consumption.

Storage tanks applied to store high viscosity derivatives like heavy crude oils, asphalt, lubricants, and paraffin are normally equipped with heating systems. Figure 4.9 presents a typical thermal heating system for storage tanks.

4.5.5 Draining or Dewatering System

Despite the efforts to reduce the water content of the derivatives in the processing units, it's practically impossible to reach a totally dry product in the final or intermediate storage tanks. The separation of oil and water phases in the separation drums occurs through the sedimentation process, due to the density gap between the water and the stored hydrocarbon. Considering that the sedimentation process can be theoretically described by the Stokes Law, according to Equation (4.1).

$$v_R = \frac{1}{18}\frac{d_p^2 g\left(\rho_p - \rho_f\right)}{\mu_f} \tag{4.1}$$

According to Equation (4.1), the sedimentation velocity is proportional to the density gap. In the case of heavier derivatives, this gap is lower, leading to a lower sedimentation velocity and the need for higher residence times for adequate separation. This information should be considered to calculate the necessary separation time between the water and hydrocarbon. Once the water has a higher density, it will accumulate in the bottom of the tank, allowing the drainage of the contaminated water to a safe location.

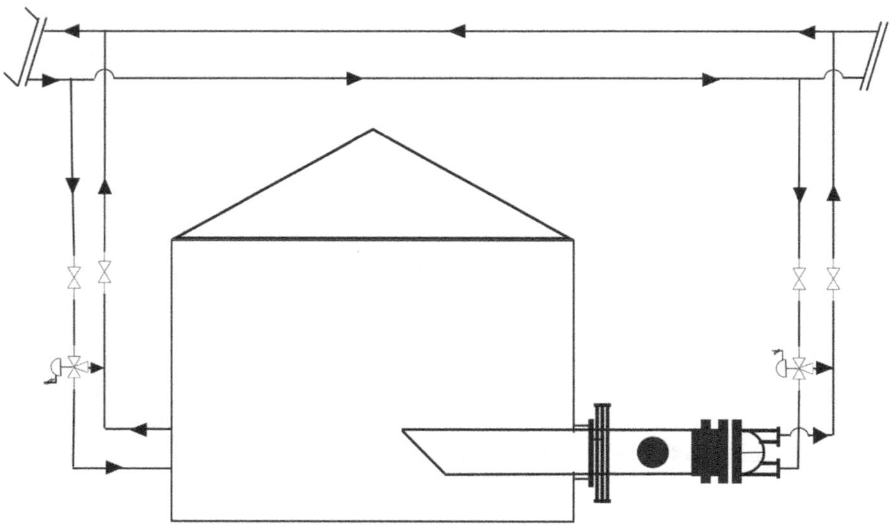

FIGURE 4.9 Typical heating system for storage tanks based on steal coil.

The draining system of hydrocarbon storage tanks is normally designed to direct the contaminated water to the wastewater treatment system of the refinery. Special attention should be dedicated to the draining system of gasoline and naphtha tanks. The water contained in these tanks normally presents high concentrations of phenolic compounds, which are toxic to the microorganisms responsible for reducing the carbon-based organic pollutants.

The water drained from gasoline and naphtha storage tanks should be directed to adequate residue tanks and fed to a conversion process capable of dealing with and destroying phenolic compounds.

The water accumulation in the storage tanks can produce a wide range of operating and integrity issues, such as microbiological degradation of the hydrocarbons producing sludge and pipeline degradation of the refinery and end-user assets. Water accumulation can also impact the quality characteristics of the stored derivatives, leading to higher operating costs. This topic tends to be a key point for refiners in the near future, considering the growing trend of renewable coprocessing in the crude oil refineries once renewable raw materials raise the tendency of water retention in the final derivatives. Due to the chemical structure, biodiesel, for example, tends to retain more than eight times more moisture than fossil diesel, which can lead to issues like microbiological degradation of the fuel in the transport and storage systems. The soluble water content in pure biodiesel can reach close to 1,800 ppm, while the value of the diesel with 20% biodiesel can reach close to 280 ppm of soluble water, and the diesel with 5% biodiesel can present until 150 ppm. This fact will lead the refiners to adequate their hardware to allow the water removal from the final derivatives, applying draining systems or the application of salt filters to control the moisture content in the final derivatives.

4.5.6 Bund

The main objective of the bund is to contain eventual leaks and minimize emergency scenarios. Normally, the crude oil refinery tanks have individual bunds, but it's possible to find tanks sharing bunds, especially in older refineries.

It's recommended to design tank farms with individual bunds once this approach tends to be safer than the design considering sharing bunds once a tank fire in a bund led to other tanks in the same bund being ignited, scaling up the emergency scenario. The size of the bunded area should be able to contain the total volume of the tank at its maximum capacity. The dimensions of the bund and the bunded area should meet the requirements of NFPA 30, aiming to ensure the security requirements in emergency scenarios.

4.5.7 Tank Farms

A well-designed crude oil refinery should be able to absorb the market fluctuations as well as operating issues with minimum impact on their operating flowrate. To ensure this goal, it's necessary that the refinery relies on an adequate tank farm, which will offer storage space and consequently time to allow the management team to take the decision-making process regarding the production planning in order to minimize the impact of operating issues of processing units over the market.

An adequate tank farm offers significant operational flexibility to the refinery, including the capacity to decide on storage derivatives waiting for an expected price improvement, keeping the processing flowrate, or keeping the operation performance due to the shutdown of key processing units.

A typical case of tank farm design is the capacity of the crude oil storage in the refinery. In the past, the refiners tended to build large crude tank farms, aiming to protect themselves from supply issues like geopolitical crises or great price variations, but this strategy demands high capital spending, and nowadays the best approach is taking into account the utilization rate of the tanks, planned outages, and the desired process flexibility to deal with the different products produced by the refinery (different grade oil products like diesel and gasoline will demand segregated tanks).

The design of tank farms should consider the necessity to stop the tanks in defined intervals to carry out internal inspection according to the asset integrity plan of the refinery. This is especially critical for the tanks which act as feed stream tanks of the processing units. The tank farm design should consider operational flexibility to allow another tank to substitute a feed stream tank to avoid impacts on the processing capacity of the refinery. Another key question in the tank farm design, from the process and operability point of view, is the consideration of residue and brine receiving tanks; again, the tank farm design needs to consider the reserve tanks for these process functions, as the unavailability of this equipment will lead to the shutdown of the refinery.

Residue tanks are used to receive processing streams during operational instability of the processing units, like shutdown and start up. These streams contain a mixture of sour water and hydrocarbons and are stored in a floating roof tank due to the high concentration of NH_3 and H_2S. The inventory of the residue tank is normally reprocessed in processing units like FCC and distillation.

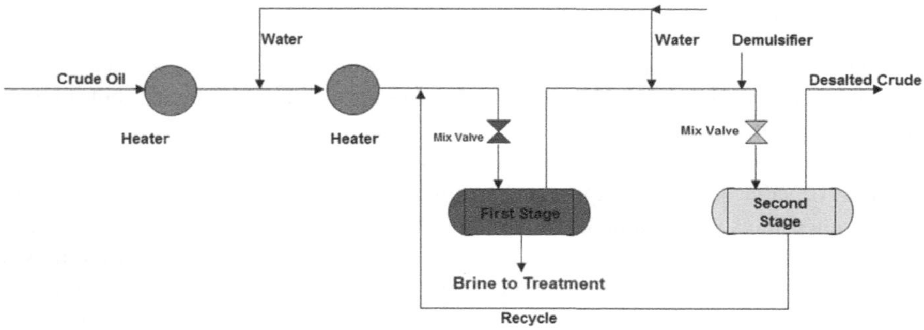

FIGURE 4.10 Crude oil desalting process with two separation stages.

The brine tank is used to store brine produced by the desalting process of the crude oil distillation unit.

The desalting process involves the mixture of crude oil with water, aiming at the dissolution of salt, considering the higher solubility of these compounds in the aqueous phase. Figure 4.10 depicts a typical desalting process with a single separation stage.

The crude oil is pumped from the storage tanks through the heating battery, where it is heated and mixed with dilution water. The mixture is made by a mixing valve that promotes an intense mixture through a pressure drop. A major part of the water is under the free form and is removed by decantation due to the difference of specific gravity between the aqueous and oil phases; however, part of the water is emulsified in the oil phase and requires actions to break the emulsion and allow the decantation of this water and the dissolved salts. This stream is called brine and is stored in a tank and then sent to the wastewater treatment unit.

Normally, the refiners choose crude oil storage tanks to act as residue and brine accumulation tanks due to their dimensions and design characteristics.

4.5.8 Blanketing System

Blanketing is a process where a seal of inert gas between the liquid level and the vapor space of the tank is provided and kept. This seal is kept under very low pressure (typically lower than 1 psig). The most common inert gas applied in the blanketing system is nitrogen and it is very common to refer to this system as nitrogen blanketing.

A blanketing system can be demanded in a storage tank design with the following objectives:

- Keep air and moisture out of the storage tank to protect the hydrocarbon integrity. Some hydrocarbons, such as delayed coking streams, would experience deterioration if in contact with air and/or moisture. Normally, the storage tanks are applied to storage feed for hydroprocessing units, where the chemical degradation of the hydrocarbon can cause fouling and a low operational lifecycle of the processing unit.

- For high-corrosive liquids, the blanketing system can be applied to prevent the internal corrosion of the tank, especially on the roof.
- Reduce the oxygen content inside the tank, aiming to keep the atmosphere in the vapor space below the flammable range.
- Minimize fugitive emissions; this is the case of tanks applied to store sour water as feed for sour water stripping units. The storage without a blanket will lead to H_2S and NH_3 emissions into the atmosphere.

4.5.9 Applying the Knowledge

During your routine operation, a stockpiling operator of a crude oil refinery called the supervisor by radio and informed them that the floating roof of a crude oil storage tank was partially sunk with around 500 mm of water level on one side of the roof. The night before, the tank had been filled with a blend of crude oil and petroleum condensate (a kind of very light crude oil). The tank was equipped with a buoyroof design floating roof. The operation of the tank was stopped, and a root cause analysis was carried out and determined that the high vapor pressure of the liquid inside the tank together with insolation on the deck roof was responsible for creating vapor under the floating roof, destabilizing the roof, and causing the roof to sink.

BIBLIOGRAPHY

API Standard 2000, 5th ed., April 1998 – Venting Atmospheric and Low-Pressure Storage Tanks Nonrefrigerated and Refrigerated.

Geyer, W. B. *Handbook of Storage Tank Systems – Codes, Regulations and Design*, CRC Press, 1st edition, 2000.

Lieberman, N.P.; Lieberman, E.T. *A Working Guide to Process Equipment*, McGraw Hill Book Company, 4th edition, 2014.

Long, B.; Gardner, B. *Guide to Storage Tanks and Equipment*, Wiley Book Company, 2004.

Myers, P.E. *Above Ground Storage Tanks*, McGraw Hill Book Company, 1st edition, 1997.

Sutton, I. *Plant Design and Operations*, Gulf Professional Publishing, 2nd edition, 2017.

5 Blending Operations (Crude Oil, LPG, Gasoline, Fuel Oil, and Diesel)

The main objective of a refining asset is to maximize the added value to the processed crude oil. One of the main ways to maximize the added value to the processed crude and improve the refining margins is to maximize the addition of low added value streams in the crude oil derivatives, respecting the quality requirements of the final derivatives.

The blending process involves the mixture of different streams, aiming to reach the final derivative with a minimum giveaway, which is the overspecification of the final derivative that represents an important way to destroy added value in a typical crude oil refinery. The blending operations are another fundamental process in a crude oil refinery which is under the responsibility of the transfer and stockpiling operations sector and is a key process both regarding the profitability and the safety of the refinery.

5.1 QUALITY REQUIREMENTS OF CRUDE OIL DERIVATIVES

Despite the recent trends indicating the reduction in transportation fuel demand, especially in markets economically developed, the current society still strongly depends on crude oil derivatives to sustain its economic and technological development.

Over the last decades, there have been great efforts in the sense of minimizing the environmental impact of crude oil derivatives. These efforts were translated mainly into severe regulations with the objective of improving the performance and reducing, in a significant way, the content of contaminants in these products, thus reducing his environmental footprint.

Crude oil processing produces a series of derivatives with distinct demands and added values. Figure 5.1 presents a simplified process flow diagram for a typical atmospheric crude oil distillation unit and the main derivatives produced in this unit.

The stream considered as fuel gas is normally composed of hydrocarbons in the range C1 to C2 and is applied as fuel in the fired heaters and boilers in their own refinery. According to the adopted refining scheme and the integration grade between refining and petrochemical operations, the fuel gas produced by thermal or catalytic conversion units (e.g., FCC or delayed coking) can be sent to petrochemical units, aiming to recover the olefins, according to the process scheme shown in Figure 5.2. In this case, the demand for fuel gas in the refinery is complemented by natural gas.

DOI: 10.1201/9781003472216-5

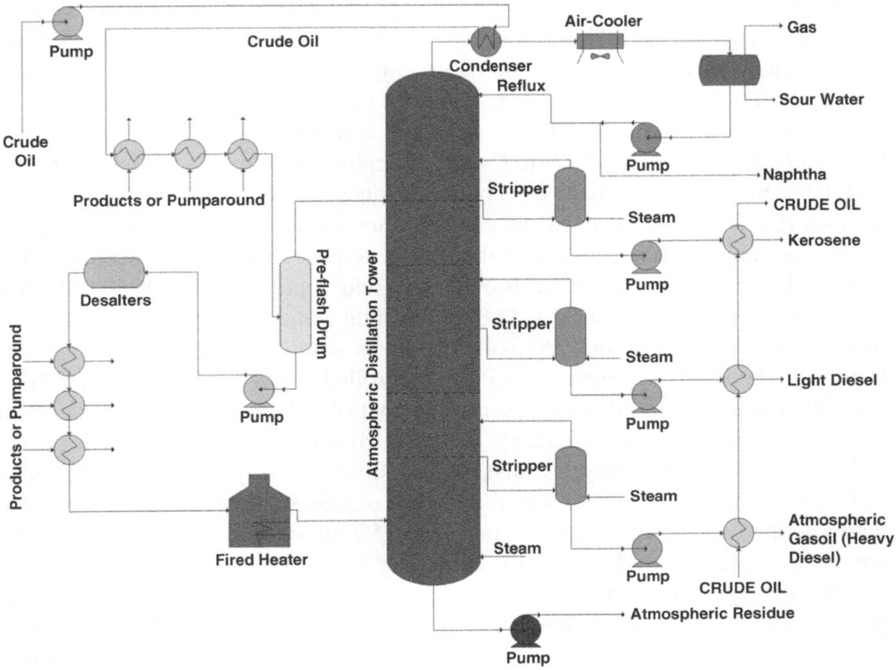

FIGURE 5.1 Typical process arrangement of an atmospheric crude oil distillation unit.

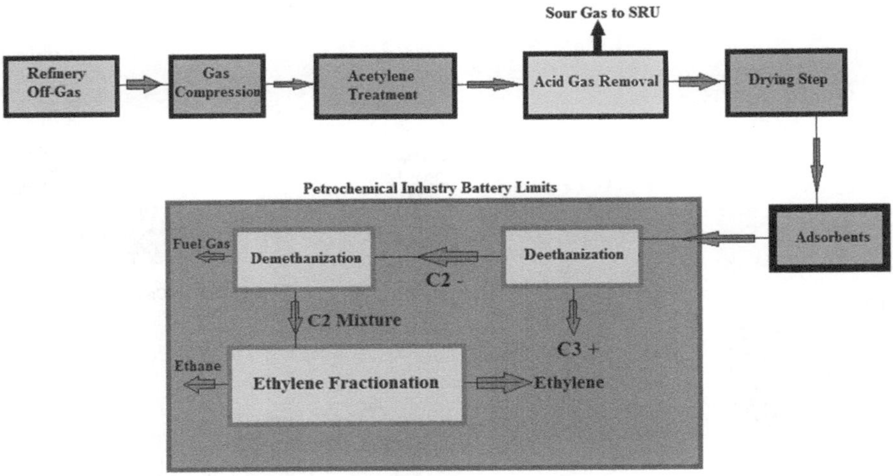

FIGURE 5.2 Block diagram of the olefins recovery process from refinery off-gases.

The main quality parameters controlled in the fuel gas are the humidity and hydrogen sulfide (H_2S) content. These requirements are normally controlled in dehydration units using propylene glycol and amine treating units, respectively. The concentration of H_2S is controlled below 1.0% in volume as well as the humidity content.

Liquefied petroleum gas (LPG) is normally composed of paraffinic and olefinic hydrocarbons in the range of C3 to C4 and is applied as domestic and transportation fuel in specific cases. The LPG can contain low quantities of light and heavy hydrocarbons (C2 and C5); however, the concentration of these compounds needs to be minimized, aiming to not prejudice the quality requirements. The concentration of light hydrocarbons is controlled through the Reid vapor pressure (RVP) which is determined by the LPG heating at 37.8°C. The light content is controlled for security reasons, aiming to keep the LPG volatility under safe values to allow storage and handling. The RVP of commercial LPG is controlled below 1430 kPa. Once LPG is normally burned in closed environments, the control of burning residue is one of the most important quality requirements of this derivative. The heavy content is controlled through a weathering test that evaluates the difficulty in the vaporization of LPG, measuring in an indirect way the content of C5+ in the mixture, which is normally defined by the boiling temperature of 95% in volume of the mixture under atmospheric pressure and is normally controlled below 2.0°C.

The other quality requirements of LPG are the sulfur content and corrosivity. The sulfur content is controlled below 140 ppm and this parameter, as well as corrosivity, is managed through desulfurization processes like regenerative caustic treatment, which is presented in Figure 5.3.

A refinery LPG pool is normally composed of the LPG produced in fluid catalytic cracking (FCC) units that are the main producers of this derivative, delayed coking units, catalytic reforming, and atmospheric distillation, and, in some cases, LPG

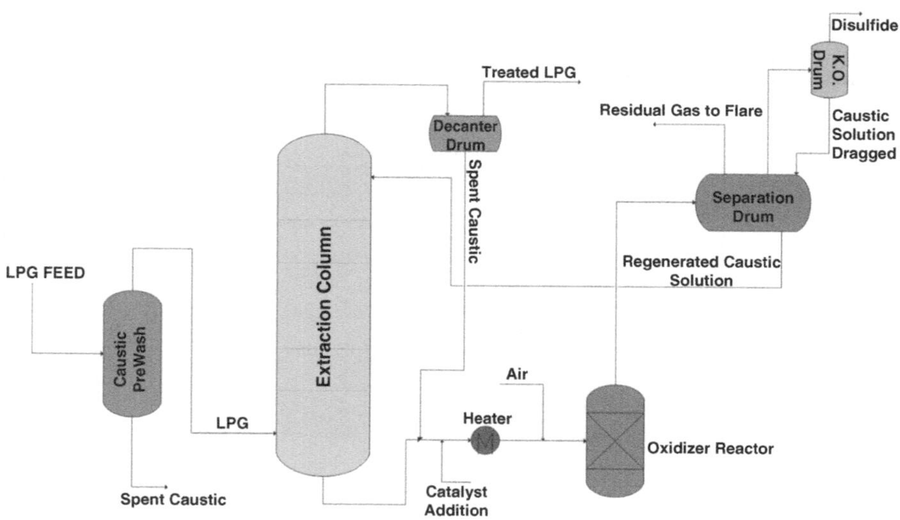

FIGURE 5.3 Typical process arrangement of LPG regenerative caustic treating.

from natural gas treating plants. The LPG from FCC can be submitted to additional separation processes to recover the propylene and direct it to the market as a petrochemical intermediate due to its higher added value when compared with the LPG.

The naphtha streams are normally directed to the refinery gasoline pool, according to the refining configuration and the demand of the market where the refiner is inserted. The streams that compose the gasoline pool also depend on the refining scheme; however, it's common for the composition of the gasoline pool to include straight-run naphtha, cracked naphtha from FCC units, reformed naphtha from catalytic reforming units, isomerized naphtha from isomerization units, and alkylated naphtha produced in catalytic alkylation units.

The gasoline is composed of the blending of these streams containing hydrocarbons with a boiling range of 30°C to 215°C (C4 to C10). Among the main quality requirements of gasoline are the antiknock capacity, volatility, corrosivity, pollutant emissions, and the tendency of combustion residue formation in the engines. The gasoline antiknock capacity is measured through the octane number that is determined by applying isooctane (2,2,4-trimethyl pentane) as a standard with an octane number of 100 and n-heptane with an octane number of 0. The octane number represents a volumetric percentage of isooctane in a mixture with n-heptane which burns with the same antiknock quality of the analyzed gasoline (measured through sound intensity). The tests to determine the octane number can be the motor octane number (MON) test and the research octane number (RON) test; a common automotive gasoline has an octane number close to 85. The naphtha streams that add higher antiknock capacity to the gasoline are the cracked naphtha from FCC units due to the high olefin content. The reformed naphtha is due to the high aromatic concentration, and the naphtha from catalytic alkylation is due to the ramified characteristics of the produced kinds of paraffin. However, the aromatic and olefin content are normally controlled in the final gasoline due to the toxicity and high volatility of these compounds.

The volatility of gasoline is related to the light content in the mixture being directly responsible for the cold starting facility of the internal combustion engines. Gasoline Reid vapor pressure (PVR) indirectly measures the amount of light present in the blend, as well as for LPG, the gasoline PVR is determined at 37.8°C (100°F) at 1.0 atm and is usually controlled below 55 kPa.

The corrosivity and emissions of gasoline are controlled through the sulfur content in the final product. Currently, the sulfur content in the gasoline is controlled below 50 ppm. For this reason, it's practically impossible to meet this specification without hydrotreating units. Selective hydrotreating units are applied mainly to treat cracked naphtha, aiming to reduce the sulfur content with a minimum loss of antiknock capacity (due to olefins saturation). The resistance of deposit formation is directly related to the olefins content in the mixture. These compounds are chemically unstable and suffer polymerization processes, forming polymers that produce deposits and inefficient combustion. The use of antioxidant additives and detergents in the final gasoline can minimize these effects.

A special case of commercialized gasoline is the aviation gasoline that is applied to airplanes equipped with Otto Cycle engines. In this case, the hydrocarbons that compose gasoline have a stricter boiling range (30°C to 170°C) containing ramified paraffin produced by catalytic alkylation process units, as presented in Figure 5.4.

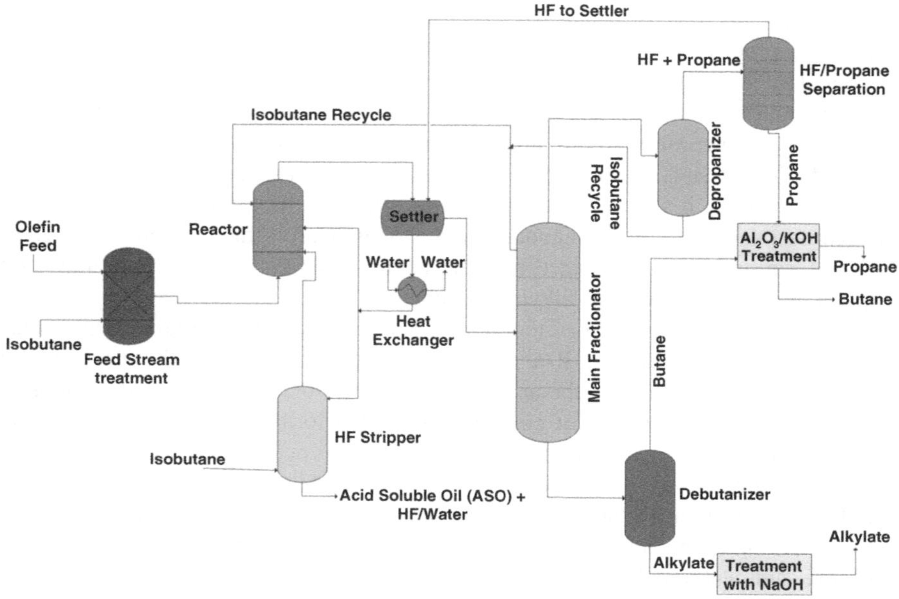

FIGURE 5.4 Typical process flow diagram of catalytic alkylation units applying HF as homogeneous catalyst.

According to the market to be supplied and the interaction level of petrochemical and refining operations, the light straight-run naphtha can be commercialized as petrochemical naphtha. In this case, it's necessary to guarantee a paraffin content higher than 65%. This alternative tends to be even more applied in the face of the tendency for a reduction of transportation fuel demand; furthermore, in markets with high demand for middle distillates, the heavy straight-run naphtha can be directed to compose the diesel or jet fuel pool.

In its turn, jet fuel is a mixture of hydrocarbons between C5 and C15 with a boiling range of 150°C–300°C and is applied as fuel to jet turbines, normally applied in aviation. Due to the severity of use conditions, jet fuel has quality requirements that are quite restricted. The combustion needs to be the cleanest possible to avoid depositions. For this reason, the polyaromatic content is controlled; this is achieved through the smoke point test.

The characteristics of flow under low temperatures are fundamental to the jet fuel, due to the operational conditions that can achieve temperatures of – 50°C. The maximum freezing point for commercial jet fuel is – 47°C. For this reason, it's fundamental to ensure an adequate cut point in the distillation step to avoid the drag of heavy paraffins to the intermediate kerosene. The thermal stability is measured through the jet fuel thermal oxidation test (JFTOT), which simulates the operational conditions that the fuel is submitted to.

The corrosivity and chemical stability in relation to the materials applied to the construction of turbines are controlled through the content of total sulfur, mercapto sulfur, and H_2S. Normally, the jet fuel is submitted to caustic treatment to control these compounds. In modern refining units, this step is carried out in hydrotreating units. The flash point (minimum 40°C) and the electric conductivity are other requirements directly related to the security in the derivative handling.

Diesel is the crude oil derivative that has had the most increasing demand in the last decades. This derivative is mainly applied as a transportation fuel by vehicles equipped with diesel cycle engines, composed of hydrocarbons between C10 and C25 with a boiling range of 150°C–380°C. The diesel ignition quality is measured through the cetane number that corresponds to a volumetric percentage of cetane (n-hexadecane) in a mixture with heptamethylnonane, which burns with the same ignition quality as the analyzed diesel. The linear paraffinic hydrocarbons are the compounds that contribute most to the diesel ignition quality, raising the cetane number, while the presence of aromatics reduces this parameter and prejudices the ignition quality. Currently, the minimum cetane number of commercial diesel is 48. In some countries, such as Brazil, the addition of biodiesel in the final product with a minimum concentration of 12% in volume is mandatory.

The diesel volatility is controlled, aiming to ensure the cold start performance and safety during the handling; the minimum flash point of 38°C and the distillation curve temperatures correspond to 50%, 85%, and 95%, and the recovered volume is controlled in determined limits to ensure the total vaporization in the working conditions. These parameters limit the quantity of naphtha added to the diesel pool.

Another important parameter controlled in the diesel is the plugging point that aims to control the content of linear paraffins that tend to crystallize under low temperatures, prejudicing the fuel supply to the engine. The plugging point is determined according to the weather conditions in the region of application. In Brazil, the plugging point is controlled in the range of 0–10°C.

Diesel emissions control is carried out by managing the fuel density, aiming to control the content of heavy compounds, especially polyaromatics. Currently, the density of commercial diesel is controlled in the range of 830–865 kg/m^3; for ultra-low sulfur diesel (ULSD), this parameter is controlled below 850 kg/m^3. In the last decades, there have been great efforts to reduce the environmental damage produced by diesel burning. Nowadays, environmental regulations require the commercialization of low sulfur diesel with a maximum sulfur content of 10 ppm, despite some markets, mainly in developing countries, still commercializing diesel with higher sulfur content (500 ppm), but this tends to change in the near future.

This requirement led to the necessity of refiners to expand their hydrotreating capacity. Figure 5.5 presents a simplified process flow diagram for a high severity hydrotreating unit.

Viscosity is also a controlled parameter in the diesel, aiming to ensure adequate nebulization in the combustion chamber. High viscosities can be prejudicial due to the poor dispersion of the fuel, while low viscosities lead to excessive dispersion. Normally, the diesel viscosity is controlled in a range of 2.0–5.0 mm^2/s. The diesel lubricity is measured to control the wear due to the friction of the pieces in contact with diesel and is determined by specific tests. The lubricity and the electric conductivity are directly related to the concentration of polar compounds that are reduced after the hydrotreating step. For ultra-low sulfur diesel, normally additives are applied to correct these parameters.

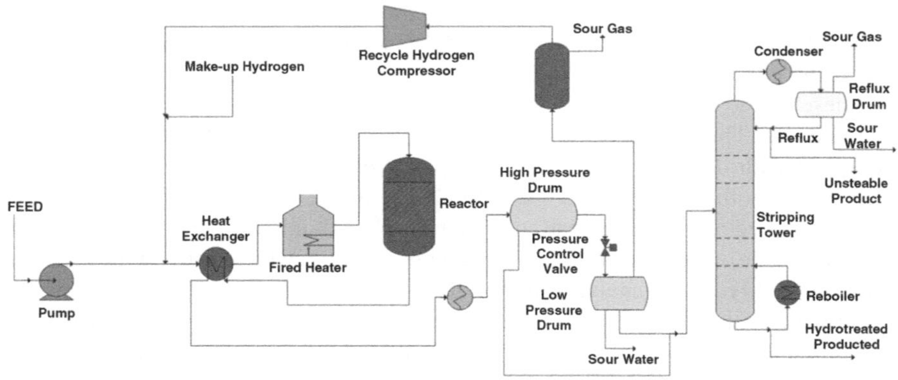

FIGURE 5.5 Basic process flow diagram of high severity hydrotreating unit.

The control of water content and sulfur, nitrogen, and aromatic compounds aims to avoid the proliferation of microorganisms that lead to the filters plugging and adding corrosivity to the derivative, as well as raise the stability to oxidation and deposit formation.

Adequate management of crude oil derivatives quality requirements is fundamental to achieving the desired goals of performance, safety, and environmental impact. Ensuring the efficiency and reliability of the process responsible for controlling these parameters is a key factor in achieving competitiveness and sustainability in the refining industry.

Light products are crude derivatives with higher demand and higher added value, being responsible for a major part of the economic result in the refining sector. For this reason, in the last decades, the refiners have been enlarging the capital investment, aiming to improve the bottom barrel conversion capacity and thus minimize the production of low added value derivatives.

Despite this fact, the so-called dark products have specific applications that are fundamental to the social and economic development of contemporary society. The industrial fuel oil is produced based on vacuum residue obtained in the vacuum tower bottom, as presented in Figure 5.6.

The fuel oil formulation is carried out by adding diluents to vacuum residue, aiming to achieve a specified viscosity according to the application. Commonly used diluents are the gas oil streams from vacuum distillation or streams from deep conversion units like FCC (light cycle oil, LCO) or delayed coking (light and heavy gas oils). In some cases, diesel is applied as a diluent. The main quality parameters controlled in fuel oil production are sulfur content, viscosity, the content of sediments and water, vanadium concentration, flash point, and pour point.

The fuel oil is considered a low sulfur fuel when the maximum concentration of this contaminant is 1.0% by mass and high sulfur fuel when the maximum sulfur concentration is 2.5%. The sulfur content control aims to impose a limit on the emissions of harmful gases during the derivative burning. The viscosity control in the fuel oil aims to minimize the transfer costs and ensure adequate flow and vaporization in the burners. The kinematic viscosity of industrial fuel oils (measured at 60°C) is controlled in the range of 600–950 mm^2/s.

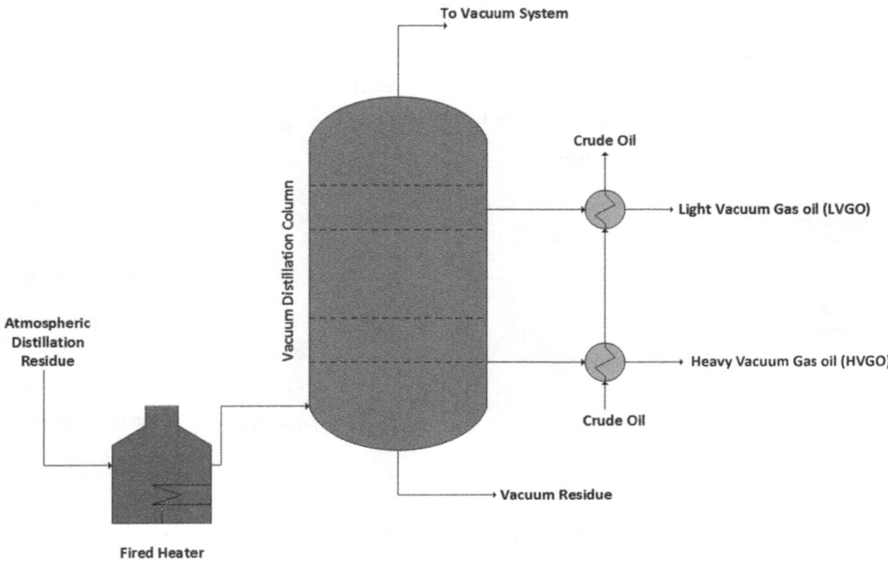

FIGURE 5.6 Vacuum crude oil distillation to transportation fuels production.

The limit of water and sediment content aims to minimize the fouling, deposition, and corrosion in the process equipment and damage to the burners. Furthermore, the water presence reduces the calorific value once part of the released energy is applied to vaporize the water and can provoke flame instability. The maximum vanadium content control aims to minimize the effects of the chemical attack of this metal on the refractory of boilers and fired heaters, as well as metallurgic damages; the maximum vanadium content in the fuel oil is 200 ppm. In its turn, the flash point is applied to control the fugitive emissions and add security during the derivative handling, while the pour point aims to ensure the flow under low temperatures. The pour point specification relies on the weather conditions in the application region.

In some cases, it's necessary to mix different fuel oils to meet the quality requirements. In these cases, it's important to consider the compatibility between the fuel oils. Oils from highly paraffinic crudes show chemical incompatibility with oils produced from crude oils with high asphaltenes content, once the presence of paraffin tends to precipitate the asphaltenes due to the resins solubilization that stabilizes the asphaltenes in the solution.

Asphalt is considered a residual fraction of crude oil, normally composed of molecules predominantly aromatic. Asphalt is produced from the vacuum residue that is obtained in the bottom of the vacuum tower, according to the aforementioned or from the dilution of the asphaltic residue obtained from the solvent deasphalting process, as presented in Figure 5.7.

The main application of asphalt is the composition of road pavements. Among the asphalt quality requirements are consistency and hardness, ductility, thermoplasticity, viscoelasticity, thermal susceptibility, and durability.

The determination of consistency and hardness of the asphalt aims to define the handling capacity of the derivative. This variable is evaluated by the penetration test,

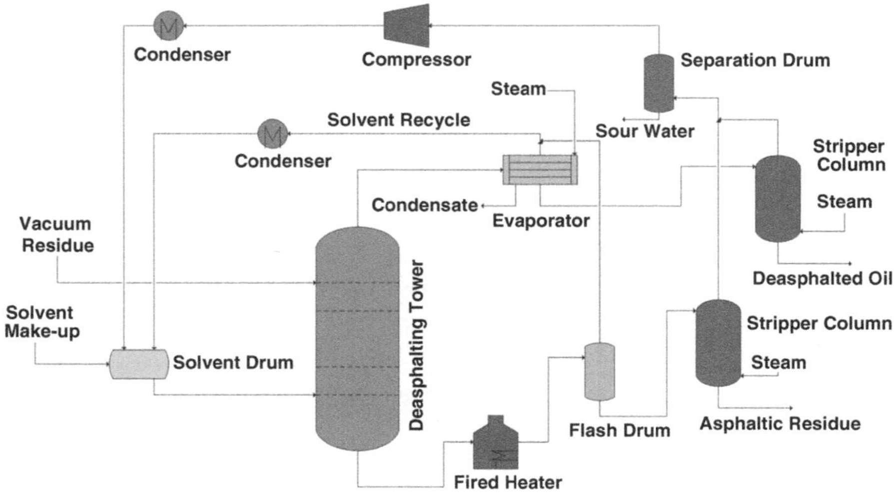

FIGURE 5.7 Conventional solvent deasphalting process.

which is performed using a standard needle under specific conditions of loading, temperature, and time. The ductility measures the ability of the asphalt to elongate before rupture; this requirement is directly linked to the strength of the material when applied to the pavement composition.

The thermoplasticity and the viscoelasticity are controlled, aiming at the possibility of hot application of the asphalt and the restoration of the properties of the material after cooling. Thermal susceptibility gives the asphalt the ability to withstand temperature variations without loss of properties such as consistency and ductility. In turn, the durability test is performed under an aggressive atmosphere of exposure to air and heat, and the other properties are subsequently re-evaluated. The asphalt flashpoint is controlled below 235°C to provide safe handling of the derivative.

The marine fuel oils, called bunkers, are produced from the bottom residue of vacuum distillation. These derivatives are applied as fuels to large ships that operate with diesel cycle engines; thus, despite being produced from vacuum residue, the bunker oils have quality requirements different and more severe than the industrial fuel oils.

Due to the application in diesel engines, it is necessary to control the ignition quality of the bunker. This requirement is evaluated indirectly through the Calculated Carbon Aromaticity Index (CCAI), which is evaluated from the density and viscosity parameters that are controlled.

Viscosity is an extremely important variable for the bunker since it is directly related to the ease of nebulization of the derivative in the combustion chamber; high viscosity oils require a higher heating rate before firing, and the bunker viscosity is generally controlled between 2.0 and 11 mm^2/s (measured at 40 °C). Another important feature is the pour point of the bunker. This variable depends on the climatic conditions in the region of application, since it is related to the capacity to flow at reduced temperatures, and the bunker pour point is normally controlled between –6.0 and 6.0°C.

The density of the commercial bunker is controlled between 877 and 897 kg/m^3, while the minimum flash point is 60°C in order to limit fugitive emissions and give safety to the handling of the product. The maximum water and sediment content for the commercialization of the bunker is 0.40% by volume in order to avoid corrosion and waste deposition in equipment and storage tanks.

Nowadays, the maximum sulfur content in the bunker is 3.5 % (m.m); however, aiming to reduce the environmental impact of this derivative, the regulation IMO 2020 establishes 0.5% (m.m) as the maximum sulfur content to the bunker after 2020.

As aforementioned, the bunker is produced from the vacuum residue; this stream concentrates a major part of contaminants in the crude oil. In this sense, to comply with the IMO 2020, the refiners can adopt some of these strategies:

- Process low sulfur crudes: In this case, the refiner would lose operational flexibility and would be forced to restrict his supplier's slate which can be a threat in case of geopolitical instabilities. Furthermore, the low sulfur crude oil cost tends to rise after 2020, putting the refining margins under pressure.
- Apply low sulfur streams as diluents: Once bunker is produced through the vacuum residue dilution with lighter streams, a possibility is to apply clean streams like diesel to produce bunker; however, this strategy evolves the use of high added value and high demand streams as diluents, which can lead the refiners to lose market and competitiveness.
- Refining hardware modification enlarging the bottom barrel conversion capacity: Despite being the strategy with the higher cost, this solution would raise the refinery capacity to convert low added value streams into derivatives with high demand like diesel and jet fuel, as well as the reduction of contaminants content in heavy streams like atmospheric residue, thus avoiding the scent of high added value streams to compose the bunker pool.

The new specification regarding bunker oil production should have significant effects on the downstream industry, such as a rise in operating costs and a consequent reduction of refining margins, and the need for increased capital investment for the installation of deep conversion and elevation of refinery complexity. In this way, refineries configured for high conversion capacity, such as the coking/hydrocracking configuration, will have a competitive advantage over less complex refineries (e.g., cracking configuration) due to higher generation of lower value streams in these process units.

On the other hand, the new legislation may represent an excellent trading opportunity for countries with easy access to low sulfur oil reserves. An example is Brazil, which has low sulfur reserves and refining facilities capable of producing bunkers within the specifications of IMO 2020 from the national crudes, which may make the country a relevant player in the supply of this derivative in the coming years.

The search for profitability enhancement in the marketing of his products is a constant in any business, and this is not different in the crude oil refining industry. Aggregating value to derivatives and reducing production costs are traditional ways to improve profitability; however, in the petroleum refining case, adding value to derivatives normally implies raising the operational costs, mainly related to power consumption, since the production of better quality derivatives requires more

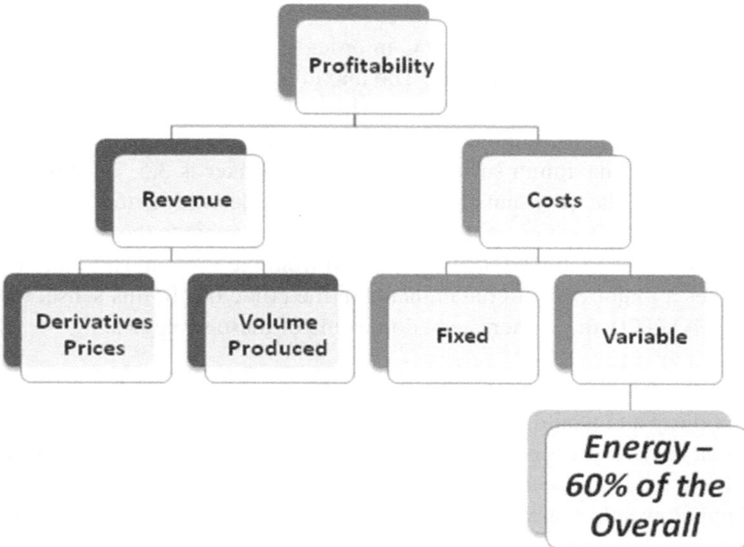

FIGURE 5.8 Simplified scheme for profitability composition in the refining industry.

complex processes. Figure 5.8 presents a simplified scheme for profitability in the downstream industry.

According to Figure 5.8, one way to raise profitability is to improve revenues through a higher production volume of derivatives and/or raise the prices of commercialized derivatives. Normally, the production volume is limited by the processing capacity of the process units and the refinery. Therefore, adding value to derivatives is the simplest way to enhance profitability in crude oil refining with minimum capital investment.

One of the most applied routes to improve profitability in the refineries is the blending process streams, aiming to add to the final products, as much as possible, low added value streams, avoiding additional costs with treatments of these streams to achieve the quality and environmental requirements (contaminants content, etc.). The hydrocarbon mixture should respect some constraints, as presented in Figure 5.9.

As presented in Figure 5.9, the hydrocarbon mixture needs to comply with restrictions, such as available component stocks, market demand, and hardware limits; keep the prices attractive; and, mainly, meet the quality and environmental requirements. The region limited by these constraints is called "blending space" and respecting these limitations, it's possible to blend low added value streams to compose final products.

The challenge of any blending system is to maximize the addition of low added value streams in the composition of final derivatives. We will describe in more detail the blending systems dedicated to preparing optimized recipes of LPG, gasoline, and diesel. Normally, the jet fuel is prepared without mixtures due to the stricter regulations related to this derivative. Despite this fact, in some refineries, it's possible to add heavy straight-run naphtha to the jet fuel pool, aiming to maximize the volume of the final derivative and improve the refining margins.

FIGURE 5.9 Schematic representation of the "Blending Space" for crude oil derivatives.

5.2 BLENDING OPERATIONS – CRUDE OIL

Choosing an adequate crude oil slate is among the most relevant decisions of refiners. Refining assets are designed considering a narrow range of characteristics of the crude oils to be processed; however, over the useful life of the assets, the crude oil slate to be processed can undergo deep changes either due to a shortage of crude oils with certain characteristics or by supply difficulties linked to geopolitical issues.

The characterization and classification of different types of crude oils aim to establish primarily their value in relation to reference crudes like Brent and WTI (West Texas Intermediate), as well as define the technological and refining routes to his adequate processing. The crude oil consists basically of a mixture of hydrocarbons and associated impurities. These impurities normally refer to sulfur, nitrogen, oxygen, and metals. The concentration of these impurities significantly raises the technological challenges of crude oil processing, leading to a reduction in crude oil prices according to the concentration of the impurities. The determination of the crude slate to be processed in a refinery is based on a blending of crude oils, aiming to achieve an adequate composition of hydrocarbons and contaminants that allow the processing in a reliable and profitable manner.

According to some scenarios, such as the discovery of abundant reserves of crude oil with characteristics different from those suitable for a given refining asset, it can support the decision of capital investments, aiming to adapt the refining assets to the processing of a certain type of crude oil. This fact is common when a refiner is an importer and oil reserves are discovered in the local market.

In relation to hydrocarbons, crude oil contains paraffinic, naphthenic, and aromatic molecules that confer the chemical and physical characteristics of crude oil.

Due to its high chemical instability, crude oil rarely has relevant concentrations of olefins. One of the main variables used for oil classification is the American Petroleum Institute (API) grade, defined based on the density of crude oil.

Normally, light oils have a higher concentration of paraffins in their composition, while medium and heavy oils tend to concentrate on naphthenic aromatic molecules in their composition. The oils considered light tend to be more commercially valued since, through direct distillation, they present higher yields in high added value chains such as diesel and aviation kerosene. In addition, the contaminants of crude oil (sulfur, nitrogen, metals, and oxygen) tend to be concentrated in the heavier fractions of oil. Thus, crude oils have a higher yield of derivatives with high added value and lower processing costs when compared to heavier oils.

The blending operation of crude oils is normally carried out in petroleum terminals, and the refiners receive the final blending by pipelines in crude oil storage tanks. Despite being an operation carried out outside the crude oil refinery, it's important to understand the concepts involved in crude oil blending once this operation directly affects the performance, profitability, and reliability of a refining asset.

The refining hardware is designed to process a specific range of crude oils, but the variation in crude oil availability, prices, and demand by crude oil derivatives tends to lead the refiners to change their processed crude oil slate. One way to maximize the refining margins is to try to reduce the cost of raw material (crude oil), and this is made by exploring the flexibility capacity of the refining hardware in order to reach a crude oil blend capable of meeting the maximum yield of high added value derivatives with minimum cost and maximum performance and reliability of the refining hardware.

The refiners try to achieve these goals by blending different types of crude oils, always looking to reduce the participation of high-cost crudes, which are normally light and sweet crudes like the Arabian light, for example. As presented in Figure 5.9, this operation should respect a series of restrictions, and, in the case of crude oil, it's important to consider the chemical compatibility of crudes. The mixture of highly paraffinic crudes with heavy crudes can lead to asphaltenes precipitation, which can lead to several operating issues in the stockpiling assets as well as in the crude oil desalting systems.

As aforementioned, the main objective of a crude oil blending operation is maximizing the addition of low-cost crudes (normally heavier and high sulfur crudes) in the final blending, aiming to reduce the total cost of the processed crude slate by the refinery and then maximize the refining margin.

The crude oil blending can be carried out through batch operations, which are made in a final tank where the inventory of different crude oils stored in the separated tanks is fed to the blending tank, as presented in Figure 5.10.

The volume of each crude to be added in the blending tank is calculated and defined by creating a document called the blending recipe. After reaching the desired volume in the final blending, the inventory of the blending tank is mixed until it reaches a homogenous mixture and then is sampled to determine if the blending recipe was successfully accomplished. If necessary, correction is carried out through the addition of the crudes to reach the desired composition and characteristics of the final blending. The main disadvantage of the batch blending operation is the high time consumption, which can lead to operating issues for the refinery, such as reduction of

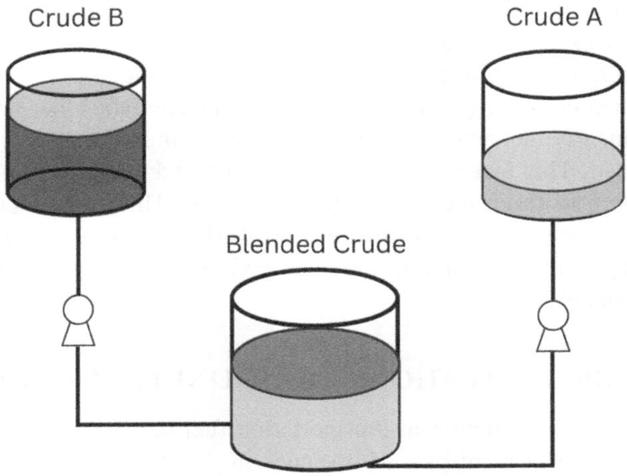

FIGURE 5.10 Batch blending operation of crude oils.

the flowrate processing of the refinery, aiming to ensure time to prepare or correct the blending of the blending tank, which will supply the distillation unit.

Considering this aspect and the technological advance in process analyzers, the refiners are investing in online blending systems, aiming to optimize the final blending and reduce the time necessary to prepare the next tank to feed the distillation unit. The online blending systems apply to a series of process analyzers to monitor the physical and chemical characteristics of the crudes and control the flowrate or proportion of each crude added to the final blending, as presented in Figure 5.11.

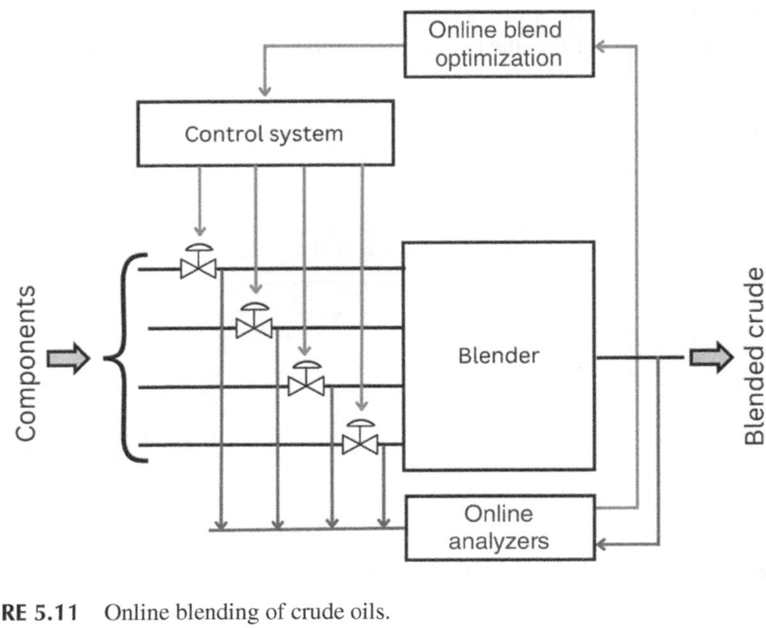

FIGURE 5.11 Online blending of crude oils.

The operation of online blending systems demands adequate reliability of the processing system, especially the process analyzers, aiming to avoid failures which will impact the operational stability of the crude oil distillation unit.

It's important considering that some quality characteristics like water and sulfur content are linearly blended while others like vapor pressure and viscosity are not linearly blended. This fact needs to be considered and adequately managed in the blending system to reach the desired final blending. This is another advantage of online blending systems: once software controls the blending system, it minimizes the probability of mistakes due to the data feed in spreadsheets by the production planning personnel.

5.3 BLENDING OPERATIONS – GASOLINE PRODUCTION

Despite the trend of reduction in transportation fuel demand, these derivatives are still responsible for a great part of revenues in the downstream industry and are fundamental to sustaining the economic development of the nations, especially in developed economies. According to Figure 5.12, the demand for gasoline tends to be sustained by developing economies in the short term.

Another example of a market with great dependence on transportation fuels is the Brazilian domestic market. The country is the world's seventh largest crude oil derivative consumer, and a major part of this consumption is related to transportation fuels, leading to the country reaching the third position in this category, demanding mainly middle distillates (diesel and kerosene), and, according to the Brazilian Petroleum Agency (ANP), the Brazilian demand for crude oil derivatives will rise 20% until 2026 against the current 2.4 million barrels per day.

Minimizing the consumption and the environmental impact of transportation fuels is a constant objective of researchers, and some refining technologies were developed over the years, aiming to improve performance and reduce the environmental footprint. Gasoline is one of the most consumed crude oil derivatives in the world and is

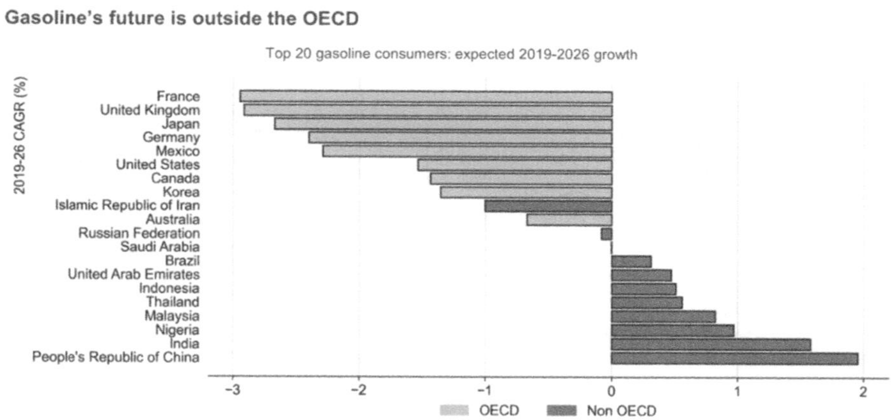

FIGURE 5.12 Growth of gasoline demand (IEA, 2021).

normally composed of a blend of naphtha, aiming to comply with strict regulations like sulfur content and octane number. An adequate octane number can optimize the performance of the gasoline in Otto cycle engines and minimize the consumption. In the refining hardware, the refiners normally rely on dedicated refining technologies to improve the octane number of the final gasoline through the management of naphtha molecules.

The streams normally involved in the gasoline production process are straight-run naphtha, cracked naphtha (after hydrodesulfurization), coke naphtha (after hydrotreatment), and reforming naphtha.

Reformed naphtha, produced in the catalytic reforming unit, is one of the streams that contribute to raising the octane number in the final gasoline. However, due to the severe restrictions related to the carcinogenic aromatic emissions, mainly benzene, some refiners have avoided the application of this stream to formulate gasoline, directing the reformed naphtha preferably to petrochemical intermediates production in aromatic complexes.

The North American market is an interesting case study which perfectly describes the challenges of producing low sulfur and high performance gasoline. Since 2020, the Tier 3 gasoline specifications put pressure on the USA refiners once they impose a maximum sulfur content in the gasoline of 10 ppm at the same time, which requires better performance (high octane numbers). In practice, these restrictions create more difficulties for the refiners to use their cracked naphtha (naphtha from FCC) in the gasoline pool once it's necessary to hydrotreat a higher amount of olefins with octane losses, leading the refiners to look for more selective catalysts and improve their hydroprocessing capacity.

As described above, the final gasoline is a blending of different naphtha streams which are available according to the refining configuration or refining scheme of the crude oil refinery.

The blending process of gasoline production involves the mixture of naphtha and butane, aiming to maximize the final gasoline volume while respecting the quality requirements and maximizing the participation of lower added value streams in the final composition.

In the gasoline case, the main objective is maximizing the volume of straight-run naphtha and butane. Butane is normally added to LPG production, which normally presents lower demand and lower added value than gasoline. Consequently, maximizing the butane participation in the gasoline pool is an important optimization goal of a typical gasoline blending system; the use of butane improves the octane of the final gasoline. Regarding the straight-run naphtha, this stream can be commercialized like a petrochemical feed with good refining margins, but in developing economies, maximizing gasoline production tends to be the main goal, and adding the maximum volume of straight-run naphtha is another objective of a typical gasoline blender.

As mentioned above, the main restriction of butane addition to the gasoline pool is the RVP, while the main restrictions of the straight-run naphtha in the gasoline pool are the sulfur content (for refineries without straight-run naphtha hydrotreaters) and the octane index. Figure 5.13 presents a typical blending system dedicated to producing gasoline.

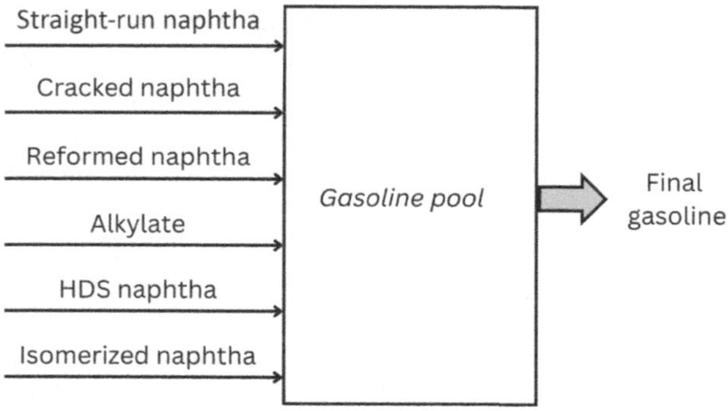

FIGURE 5.13 Example of process streams blending to produce gasoline.

Cracked naphtha (naphtha from FCC) contributes positively to the octane number of the final gasoline; however, due to the current restrictions related to the sulfur content in the gasoline (a maximum of 10 ppm), the use of cracked naphtha without a treatment step is limited. Refineries that have catalytic alkylation units in their refining scheme normally direct this stream to produce aviation gasoline that has higher market value when compared with automotive gasoline. For this reason, the participation of the alkylation naphtha is minimized in the composition of this fuel.

Isomerized naphtha has low contaminant content (sulfur and nitrogen) and a high octane number, and for these reasons, the participation of this stream in the formulation of gasoline is maximized in refineries that have isomerization units in the refining scheme. In markets with high demand for gasoline, refiners can add butanes to the gasoline pool; however, the participation is limited due to the high vapor pressure of this stream that can lead to breaking quality requirements (RVP). Normally, butanes are added to the LPG pool, respecting the limits to avoid breaking the mixture quality related to heavy limitations.

Nowadays, due to the environmental and quality regulations over gasoline, the hydrodesulfurization of cracked naphtha is fundamental to make it possible for the refiners to meet the sulfur content in the final gasoline that can be lower than 10 ppm in most restrictive markets. The blending process can be carried out in the final storage tank, calculating the volume of each stream that should compose the final gasoline recipe; the volume is calculated manually using spreadsheets and software where the restrictions related to the quality requirements are fed. Despite possible and profitable to operate under this regime, the best way to operate a blending system is through the use of an online blending system, which is depicted in Figure 5.14.

In online blending systems, the streams are fed to the system, and a group of analyzers are applied to monitor in real time the quality requirements of the final gasoline. This control system acts over an optimization control system to raise or reduce the participation of each stream, aiming to maximize the added value of the final derivative with a minimum giveaway. After reaching the total volume of the final gasoline storage tank, the mixture is blended again through the tank mixers, and an inventory sample is taken and sent to the quality laboratory for inventory certification, and then, prepare the inventory for the expedition.

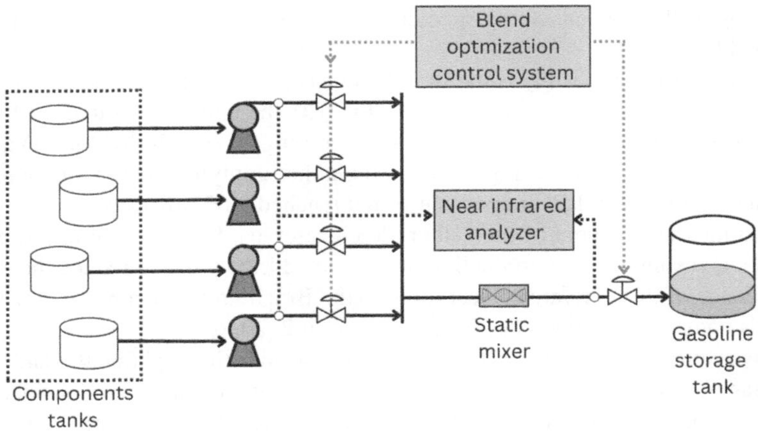

FIGURE 5.14 Example of online blending system to produce gasoline.

5.3.1 Alkylated Naphtha – A Special and Relevant Case

In June 2019, a process accident occurred at Philadelphia Energy Solutions' refinery. This incident restarted the discussion about the security of naphtha alkylation units based on hydrofluoric acid (HF) as a homogeneous catalyst. In the past few years, some refiners have decided to stop the operation of HF alkylation units due to the unacceptable safety risks.

The high toxicity and corrosion risks associated with HF require a sizable amount of attention from inspection equipment teams to control the integrity of processed equipment. When this management system fails, the consequences can be dire.

Over the last 20 years, the following process accidents involving alkylation units have brought significant attention regarding the safety characteristics of these process units:

Ciniza's oil refinery (US), 2004
Hube Globe Co.'s plant (South Korea), 2008
CITGO's Corpus Christi, Texas, refinery (US), 2009
ExxonMobil's Torrance, California, refinery (US), 2015
Husky Superior's refinery (US), 2018
Philadelphia Energy Solutions' refinery (US), 2019.

This short list of process accidents has sounded like a warning to authorities of the significant safety risks associated with the presence of HF alkylation process units near urban centers. The release of HF can affect the population and cause severe damage and loss of life. According to the US Chemical Safety and Hazard Investigation Board, 46 HF naphtha alkylation units are in operation in the United States.

It should be noted that naphtha alkylation technology as an octane-boosting route can be attractive to refiners in markets with high gasoline demand and availability of LPG, such as in the United States. This chapter provides information on available technologies that can ensure an adequate balance between process safety and gasoline quality.

An interesting case study is the Brazilian market, which has historically relied on imports to satisfy domestic LPG demand. In 2020, the internal production of LPG in Brazil was 9.86 MMm^3/yr, with refineries supplying 74% (7.34 MMm^3/yr) and petrochemical plants and natural gas processing units supplying the remaining percentage. In the same year, LPG imports in Brazil reached 3.62 MMm^3/yr.

With the development of pre-salt crude oil reserves, Brazil's natural gas production is increasing, and the production of associated LPG tends to reduce the external dependence of this derivative. According to the Brazilian Petroleum Agency (ANP), Brazil's natural gas imports decreased by 20% in 2020.

This scenario has led Brazilian refiners to consider the naphtha alkylation route for octane boosting to produce low sulfur and high-quality gasoline. Considering the growing propylene gap, another attractive route to monetize LPG surplus is building propane dehydrogenation (PDH) units.

Despite the risks related to the operation of naphtha alkylation units, alkylate is an important octane booster for gasoline. According to GlobalData, global alkylation capacity reached nearly 2.7 MM bpd in 2023 and is forecast to increase 2%/yr to 2028.

Although the United States is the leader in total naphtha alkylation capacity, China represents the strongest growth in new capacity builds.

Once the final gasoline is produced by blending different streams, the clean octane boosters like alkylated naphtha have great relevance to reach a cleaner and high performance gasoline. The use of straight-run and reformed naphtha is normally minimized to direct these streams to the petrochemical intermediates market due to the higher added value of these streams.

Cracked naphtha [naphtha from fluid catalytic cracking units (FCCUs)] helps increase gasoline's octane number during processing. However, due to restrictions related to sulfur content in gasoline [a maximum of 10 parts per million (ppm)], the use of cracked naphtha without a treatment step is limited. Refineries that have catalytic alkylation units in their refining scheme normally direct this stream to produce aviation gasoline, which has a higher market value when compared with automotive gasoline. For this reason, the amount of alkylation naphtha is minimized in the composition of this fuel.

Isomerization naphtha has a low contaminant content (sulfur and nitrogen) and a high octane number. For these reasons, the amount of this stream in the formulation of gasoline is maximized in refineries that have isomerization units in their refining scheme. In markets with high gasoline demand, refiners can add butanes to the gasoline pool; however, the amount is limited due to the high vapor pressure of this stream. Normally, butanes are added to the LPG pool, respecting the limits to avoid breaking the mixture's quality.

Due to the environmental and quality regulations of gasoline, the hydrodesulfurization of cracked naphtha can be utilized to adhere to sulfur content maximums in produced gasoline – this can be as low as 10 ppm in many markets around the world. The alkylated naphtha is a primary octane booster in the gasoline pool and is a fundamental stream for refiners to achieve specification requirements.

Traditional technological routes to naphtha alkylation. Gasoline is one of the most consumed crude oil derivatives. It is normally produced through a mixture of naphtha from different refining processing steps. The streams normally involved in gasoline production are straight-run naphtha, cracked naphtha, coke naphtha (after hydrotreatment), and reforming naphtha.

One of the main parameters of gasoline quality is the octane number, which is a measure of combustion quality. One of the streams that contribute to raising the octane number is the reformed naphtha produced in the catalytic reforming unit. However, due to stringent restrictions on carcinogenic aromatic emissions (primarily benzene), some refiners have avoided the application of this stream to formulate gasoline, directing the reformed naphtha to the production of petrochemical intermediates in aromatics complexes.

An alternative to reforming naphtha is the production of branched hydrocarbons (with high octane numbers) through the catalytic alkylation process, which involves the reactions between light olefins (C3–C5) and isoparaffinic hydrocarbons like isobutane. The reaction product – called alkylate – is a mixture of branched hydrocarbons with a high molecular weight and a high octane number. An example of a typical alkylation reaction is shown in Eq. 5.1:

$$\text{C4H10} + \text{C3H6} \rightarrow \text{C7H16}\left(2,3\,\text{Dimethylpentane}\right) \tag{5.1}$$

The reaction is catalyzed in a strongly acidic reaction environment. The acids normally employed in industrial-scale technologies are HF and sulfuric acid (H_2SO_4). The primary advantages of the alkylation process are high octane numbers, high levels of chemical stability, and streams nearly free of contaminants such as nitrogen and sulfur. These characteristics make alkylate an attractive component to gasoline formulations for the automotive and aviation industries.

Alkylation feed is generally obtained from LPG produced in deep conversion units, mainly the FCCU and the delayed coker. The LPG produced in these processing units has a high olefins content, which is ideal for the alkylation process. The isobutane stream is normally obtained through the separation of LPG produced in the atmospheric distillation unit, the FCCU, or the delayed coker in the deisobutanizer towers.

The acids generally employed as the homogeneous catalysts in the alkylation process are HF and H_2SO_4. The feed stream travels through a pretreatment step (generally molecular sieves or alumina) before being pumped to the reactor. The objective is to remove process contaminants such as water (H_2O), diolefins, and sulfur and nitrogen compounds. H_2O is especially damaging to the process, as it accelerates corrosion in piping and equipment.

After pretreatment, the hydrocarbon streams are put in contact with the HF in the reactor, and the hydrocarbon mixture and HF solution are separated through gravity in a settler vessel. The hydrocarbon phase is sent to the fractionating section, while the aqueous phase (containing most of the HF) is cooled and sent back to the reactor. As alkylation reactions are exothermic, the reactor is continuously refrigerated to maintain ideal conditions.

A portion of the HF is sent to the stripping column, where the acid is removed with isobutane. The top product is a mixture of HF and isobutene, which is sent back to the reactor. The bottom stream contains an azeotropic mixture of H_2O and HF. This step keeps the HF free of contaminants.

After the separation columns, the butane and propane streams are treated with alumina to decompose organic fluorides with potassium hydroxide (KOH) to neutralize the remaining acidity. The alkylate stream is treated with sodium hydroxide (NaOH) to neutralize the remaining acidity.

The alkylate stream is normally directed to the refinery's gasoline pool to produce high octane automobile gasoline or aviation gasoline; however, in petrochemical plants, this stream can be used as an intermediate to produce ethylbenzene (to produce styrene), isopropylbenzene (to produce phenol and acetone), and dodecylbenzene (used to produce detergents). Propane and butane streams can be sent to the refinery's LPG pool or commercialized separately.

The alkylation process using H_2SO_4 as a catalyst is similar to the HF process. However, the H_2SO_4 regeneration step is more complex and involves the decomposition of H_2SO_4 into SO_2 and SO_3 and the subsequent condensation of concentrated H_2SO_4. This regeneration can be conducted onsite or in an external processing plant. Consequently, the H_2SO_4 consumption in the process is much higher than HF. Furthermore, the solubility of H_2SO_4 in hydrocarbons is lower, requiring greater agitation to maintain the contact between the phases.

Widely used alkylation technologies are Elessent Clean Technologies' STRATCO™ Effluent Refrigerated Alkylation™ process, ExxonMobil Chemical's and Axens' ALKEMAX™ process, and Lummus Technology's CDAlky™ process. The olefins feed stream goes to a coalescer to remove H_2O. The isobutane recycling mixture is sent to the reactor. The mixture of hydrocarbons and acid heads to a settler where the phase separations occur. The organic phase is sent back to the reactor, while a control valve provides the necessary pressure reduction to vaporize the lighter hydrocarbons and remove heat from the reactor. This helps control the equipment's temperature, which increases from the exothermic reaction of the alkylation process.

The hydrocarbons are sent to a flash drum, where the lighter phase is directed to a compressor to condense in an accumulator vessel, and the propane is recovered in the depropanizer tower, while the heavier hydrocarbons (essentially isobutane) are recycled to the reactor. The stream containing the alkylate is directed to a caustic treatment and then to a deisobutanizer column where the alkylate is removed.

As mentioned, the need for catalyst replacement is higher in the H_2SO_4 process; however, the HF process needs a higher isobutane/olefins ratio, which means a greater separation system. Over the past few decades, refiners have opted to use HF alkylation technologies due to the greater simplicity of the process. It also requires less of a need for catalyst replacement, and this results in lower operational costs.

However, regulatory pressures have led some refiners to convert their HF alkylation units to operate with H_2SO_4. Due to the high volatility and higher risks presented by HF, some licensors have developed technologies to convert HF units to H_2SO_4 units. The principal process variables of the alkylation process are the isobutane/olefins ratio, reaction temperature, acid/hydrocarbon ratio, acid purity, residence time in the settler, and operational pressure.

Alkylation processes using HF have reaction temperatures from 20°C–40°C, while H_2SO_4 processes operate at lower temperatures, for example, 4°C–10°C. At higher temperatures, H_2SO_4 can suffer decomposition to SO_2 and SO_3. The operating pressure is generally sufficient to keep the hydrocarbons in the liquid phase (normally < 5 bar). The residence time in the settler is important, as not enough time can lead to undesirable reactions and the production of organic fluorides. With HF, catalyst consumption is increased and alkylate production is decreased.

Acid purity must be maintained as high as possible through the removal of acid-soluble oil (ASO), H_2O and dissolved reactants in the HF case, and through fresh acid replacement in processes with H_2SO_4 as the catalyst.

The main disadvantages of the alkylation processes with homogeneous catalysts (HF or H_2SO_4) include greater process safety risks and higher maintenance costs (primarily related to corrosion in piping and equipment).

Alternatives and safer technologies to naphtha alkylation. With an aim to eliminate these risks, some licensors have dedicated their efforts to developing heterogeneous catalysts that can replace the strong acids used in alkylate production processes (e.g., Honeywell UOP, Lurgi, Süd-Chemie, and Topsoe).

One attractive alkylation technology is Honeywell UOP's ISOALKY™ process, which is based on an ionic liquid catalyst. The heterogeneous catalyst applied in the process – in substitution of liquid acids – is based on zeolites and can offer higher operational safety for refiners and adjacent communities.

Lummus Technology has developed a solid catalyst naphtha alkylation process called AlkyClean®. The process was developed by CBI-Lummus and Albemarle. K-SAAT™ – developed by KBR – is another commercial naphtha alkylation technology based on a solid catalyst.

Despite the effectiveness of naphtha alkylation processes based on strong acids, the sizable number of incidents and accidents involving these units has forced governments and regulators to increase pressure on refiners using these units. As previously noted, there are viable technical alternatives to producing alkylated naphtha in a safe manner.

Revamping brownfield naphtha alkylation units from HF or H_2SO_4 conventional homogeneous catalysts to solid catalysts can be an attractive area for capital investments. For example, the United States has more than 45 naphtha alkylation units that could ultimately be converted. Alkylates have a significant market share in North America's gasoline pool. With the need to produce low sulfur and high octane gasoline, alkylation processing units are imperative to comply with high-quality, clean gasoline.

The catalytic alkylation process is very attractive to countries with high gasoline consumption and a sizable availability of LPG. This process can produce high octane gasoline with low levels of contaminants, although there is high capital investment and operational costs associated with this technology.

Despite the risks associated with handling strong acids like HF and H_2SO_4, naphtha alkylation units are necessary in many markets to satisfy domestic gasoline demand. It is important to note that HF and H_2SO_4 naphtha alkylation processing units must be monitored closely to ensure the integrity of these critical pieces of equipment. The failure of these units can be catastrophic, as detailed in the Philadelphia Energy Solutions' incident in 2019 and other refinery cases mentioned in the opening of this article. However, when monitored correctly, new naphtha

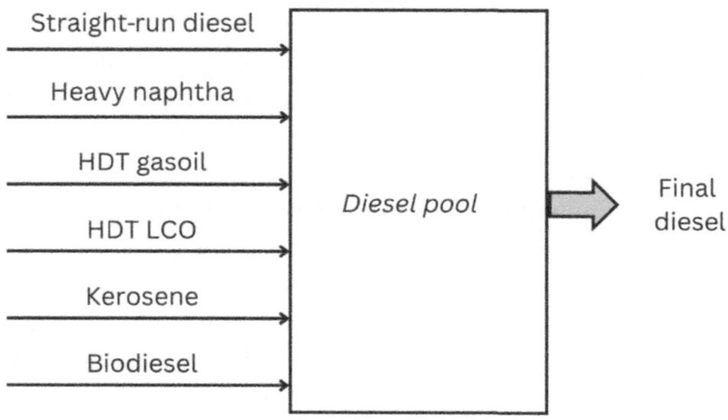

FIGURE 5.15 Example of process streams blending to produce diesel.

alkylation technologies can ensure an adequate balance between gasoline quality and process safety requirements.

5.4 BLENDING OPERATIONS – DIESEL PRODUCTION

As well as presented in the gasoline case, the blending process of diesel aims to optimize the final mixture in order to ensure maximum participation of low added value streams in the diesel pool. The streams directed to the diesel pool are exemplified in Figure 5.15.

In the diesel case, the main quality requirements that need to be respected are cetane index, flash point, plugging point, and sulfur content. Normally, the refiners tend to maximize the participation of heavy naphtha in the diesel pool, aiming to raise the production volume of this derivative. However, the naphtha presents a low cetane index and high sulfur content and raises the flash point of the final product; these characteristics limit its use to compose the final diesel.

Straight-run diesel meets a major part of the quality requirements for commercialization as diesel but has high sulfur content to meet the current specifications (sulfur lower than 10 ppm), which leads to hydrotreating necessity. The LCO produced by the FCC unit is sent to the diesel pool after the hydrotreating step but contributes negatively to the cetane index due to the high aromatics content. The use of kerosene to produce diesel is not common, since this stream is normally applied to produce jet fuel; however, his participation can improve the plugging point and cold start characteristics of the final product.

Refineries with high conversion capacity normally have thermal or catalytic conversion units that produce streams which can be directed to the diesel pool after hydrotreatment steps.

5.5 BLENDING OPERATIONS – LPG PRODUCTION

LPG production also involves the mixture of different process streams, despite being restricted processes due to the severe requirements of the final product. The LPG blending process is shown in Figure 5.16.

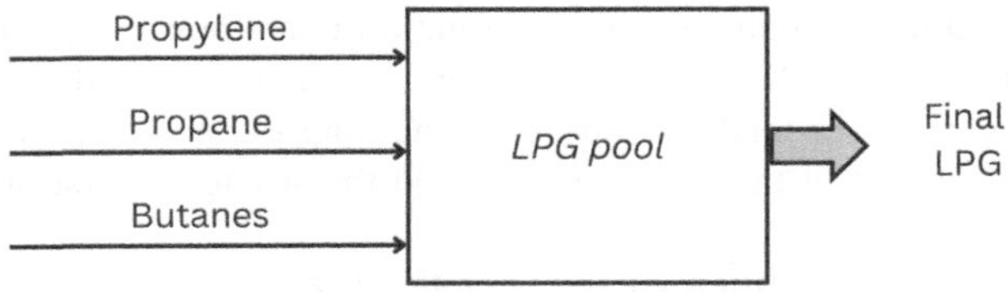

FIGURE 5.16 Example of process streams blending to produce LPG.

The light compounds content is controlled through the mixture RVP, while the heavy compounds content is controlled through the weathering; a higher weathering indicates a higher difficulty of mixture vaporization.

As described, the use of butanes to compose the LPG pool tends to be minimized, facing the price gap between the gasoline and LPG. For this reason, the refiners tend to maximize the participation of butanes in the gasoline pool.

In markets with high demand by petrochemical intermediates, the use of propylene to compose the LPG pool is minimized due to the higher added value of this stream in the petrochemical industry. Propylene is normally produced in three commercial grades: refinery grade, with purity varying from 50% to 70%; the refinery grade is applied in the production of cumene, for example. The chemical grade (92% to 96%) was applied to produce acrylonitrile, propylene oxide, and acrylic acid, and finally, the polymer grade (up to 99.5%) was applied to produce polypropylene.

The main sources of propylene are the steam cracking process. The FCC units in crude oil refineries, olefins metathesis, propane dehydrogenation, and methanol to olefins processes.

The steam cracking units are the main propylene supplier, followed by the FCC units in crude oil refineries.

According to some recent forecasts, the petrochemical market tends to rise in the next few years and, in the middle term, will be responsible for a major part of the crude oil consumption, surpassing the transportation fuels. This fact has made the refiners look for closer integration with petrochemical assets through the maximization of petrochemical intermediates in their refining hardware as a strategy to ensure better refining margins and higher value addition to the crude oil.

Considering the significant advantages of the propylene market, some refiners can consider the installation of propane dehydrogenation units to convert propane into propylene and improve, even more, the refining margins. This decision requires a deep analysis of the LPG market; once it reduces the availability of the derivative, markets with high availability of LPG, such as the United States, tend to follow this route to maximize propylene production or use the LPG stream as feed to alkylation processing units to produce cleaner and high octane gasoline.

5.6 BLENDING OPERATIONS – FUEL OIL PRODUCTION

Most of the downstream players try to minimize the production of heavy derivatives like fuel oil and asphalt due to their lower added value in comparison with light derivatives like diesel, gasoline, and petrochemical intermediates. Furthermore, these compounds tend to concentrate most of crude oil contaminants, such as sulfur and nitrogen, which produce SOx and NOx when used as fuel, and the commercialization of these derivatives tends to reduce considering the necessity to minimize the environmental impact of the energy matrix and the stricter regulations imposed on the crude oil derivatives.

One of the most impactful regulations on the downstream industry in the current and short-term scenario is the necessity to reduce the sulfur content in maritime fuels, known as IMO 2020. This regulation established the maximum sulfur content in maritime transport fuel oil (Bunker) at 0.5% (m.m) against the previous 3.5% (m.m). The main objective is to reduce the SOx emissions from the maritime fleet, significantly decreasing the environmental impact of this business.

Maritime fuel oil, known as bunker, is a relatively low-viscosity fuel oil applied in diesel cycle engines to ship movement. Before 2020, the bunker was produced through the blending of residual streams as vacuum residue and deasphalted oil with dilutants like heavy gasoil and LCO. Due to the new regulation, a major part of the refiners will not be capable of producing low sulfur bunker through simple blending.

Due to being produced from residual streams with high molecular weight, there is a tendency for contaminant accumulation (sulfur, nitrogen, and metals) in the bunker. This fact makes it difficult to meet the new regulation without additional treatment steps, which should lead to an increase in the production cost of this derivative and the necessity for modifications in the refining schemes of some refineries. Figure 5.17 presents a schematic diagram of how the bunker can be produced.

The available streams to compose the fuel oil pool depend on the reefing scheme of the crude oil refinery, but the most applied streams are decanted oil from FCC, gas oils from delayed coking and distillation units, deasphalted oil, and vacuum residue.

The drastic reduction of sulfur content in the final product led refiners to look for alternatives to reduce the sulfur content in the intermediate streams.

One of the best ways to produce very low-sulfur fuel oil (VLSFO) is blending the vacuum residue with low-sulfur diesel as a dilutant; again, an adequate blending system is fundamental to ensure reliable and profitable operations. The capacity to produce the new bunker (VLSFO) with no capital spending in deep conversion units like hydrocracking and severe hydroprocessing means a great competitive advantage, which represents lower operating costs and agility to meet the market demand.

Despite being an attractive production route, this is economically attractive only for refiners processing medium API grade and low sulfur crudes once the required dilutant volumes are relatively low. Refiners processing high sulfur crudes, like the Middle Eastern players, should apply residue hydrodesulfurization units to reduce

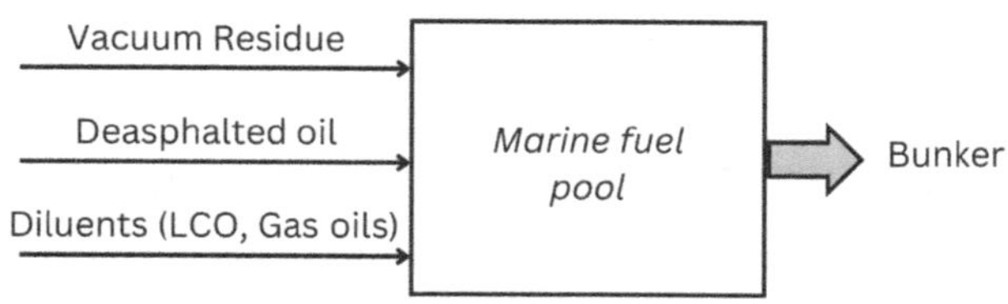

FIGURE 5.17 Bunker production process through blending process.

the sulfur content of the vacuum residue and then reach the production of VLSFO in a competitive way.

The role of the atmospheric residue hydrodesulfurization unit in the refinery goes beyond, allowing the production of low sulfur fuel oil (VLSFO). In high-complexity refineries, the unit is applied as a feedstock treatment step to conversion units such as FCC/RFCC, hydrocracking, and delayed coking. The reduction of contaminants content and residual carbon promoted by the atmospheric residue hydrodesulfurization unit significantly raises the quality of derivatives produced by downstream units as well as raises the catalyst lifecycle of deep conversion processes, such as FCC and hydrocracking, contributing to reducing the operation costs.

The process conditions tend to be more severe in the case of atmospheric residue hydroprocessing. The feedstock characteristics lead to a strong tendency of coke deposition over the catalyst, requiring higher hydrogen partial pressure (up to 180 bar in fixed bed reactors) as well as higher temperatures (380–420°C).

The hydrotreating process of atmospheric residue is normally conducted in fixed bed reactors, and the most employed catalysts are cobalt (Co), nickel (Ni), molybdenum (Mo), and tungsten (W), normally in association between them and supported over alumina (Al_2O_3). The combination Co/Mo is normally more active in hydrodesulfurization reactions, while the Ni/Mo combination is responsible for hydrodenitrogenation and aromatic saturation reactions.

A typical atmospheric residue hydrodesulfurization unit can achieve 95% conversion in hydrodesulfurization reactions and 98% in hydrodemetallization reactions. Furthermore, it's possible to achieve a reduction of 65% in residual carbon according to the employed technology. Normally, atmospheric hydrodesulfurization units rely on catalytic beds focused on removing metals, also called guard beds, aiming to protect the catalysts in the downstream reactors and improve the operational lifecycle.

Due to the severe operating conditions, the operation costs of atmospheric residue desulphurization units are higher when compared with hydrotreating units dedicated to processing distillates (diesel, jet fuel, and naphtha). The most intense hydrogenation process leads to a necessity of more robust quenching systems of catalytic beds, higher hydrogen make-up rates, and more complex phase separation systems (multiple stages).

Despite the relatively high capital cost, the implementation of atmospheric residue hydrodesulfurization units allows refiners greater operational flexibility and high sulfur oil producers greater value addition to crude oil. Countries with high production of high sulfur oil, such as Kuwait, have made a great capital investment to install atmospheric residue hydrodesulfurization units to add even more value to their crude oil reserves as well as to ensure the production of low sulfur fuel oil. Some new refineries started up recently in the Middle East rely on the biggest atmospheric residue hydrodesulfurization units in the world, aiming to ensure market compliance and high added value to the high sulfur crude oil reserves even in the IMO 2020 scenario.

The capacity to add value to the bottom barrel streams is a competitive differential in the refining industry, and this differential tends to be even more relevant in the market scenario with the IMO 2020, especially in markets with easy access to high-sulfur crude oils. Under these conditions, as aforementioned, the presence of atmospheric residue hydrodesulfurization units in the refining scheme can ensure a considerable competitive advantage for refiners.

It's important to note that the IMO 2020 creates the necessity for refiners to segregate the fuel oil tank farm to accommodate high and low sulfur fuel oils. This is a relatively common scenario in the downstream industry in which a new regulation demands significant changes in the way stockpiling assets operate, aiming to allow the refiners to deal with different derivatives specifications without contamination risks. This is another reason why a well-designed, reliable, and available tank farm is fundamental to the profitability of the downstream business.

BIBLIOGRAPHY

Gary, J. H.; Handwerk, G. E. *Petroleum Refining – Technology and Economics*. 4th ed. Marcel Dekker, 2001.

GlobalData, "Refinery alkylation units capacity and capital expenditure outlook and forecast with details of all operating and planned alkylation units to 2028," January 2, 2024, online: https://www.globaldata.com/store/report/oil-and-gas-refinery-alkylation-units-market-analysis/

Hoekstra, G. *Gasoline Desulfurization for Tier 3 Compliance*. 1st ed. Lulu, 2020.

International Energy Agency (IEA). *Oil Outlook*, 2021.

Myers, R.A. *Handbook of Petroleum Refining Processes*. 3rd ed. McGraw-Hill, 2004.

Oyekan, S.O. *Catalytic Naphtha Reforming Process*. 1st ed. CRC Press, 2019.

Robinson, P.R.; Hsu, C.S. *Handbook of Petroleum Technology*. 1st ed. Springer, 2017.

Shahnovsky, G.; Cohen, T.; Mcmurray, R. *Advanced Solutions for Efficient Crude Blending*. PTQ Magazine, 2014.

Speight, J.G. *Heavy and Extra-Heavy Oil Upgrading Technologies*. 1st ed. Elsevier Press, 2013.

Zhou, T.; Baars, F. *Catalytic Reforming Options and Practices*. PTQ Magazine, 2010.

6 Flaring Systems and How to Manage these Systems Adequately

The downstream industry is dealing with increasing demand for reducing their environmental impacts, and the players are actively called to improve their ESG policies, aiming to not only minimize their environmental footprint but also minimize their impact over the adjacent communities of the refining and petrochemical assets.

One of the most important assets in a crude oil refinery which has a great impact on the adjacent communities is the flare system. These assets are dedicated to operating like a safety system of the refinery capable of burning, in a safe way, the gases from safety relief systems of the refining processing units, especially when there is operational instability or during planned shutdown events. In other words, the flare systems are safety equipment and are essential for the refining and petrochemical assets to ensure the stability and safety shutdown of the processing units. The absence of these systems can lead to severe process accidents with great impacts on the environment and people. Unfortunately, a good example is the Bhopal disaster that occurred in a process plant operated by Union Carbide in 1984, where around 2,300 people died due to the release of a cloud of methyl isocyanate without burn once the flaring system was out of service.

Nowadays, the evolution of processing technologies leads to reliable and efficient flaring systems, but the major concern related to this equipment is to reduce the unnecessary gas flaring due to carbon emissions. According to the IEA (International Energy Agency), in 2022, the global flaring systems were responsible for the emissions of 500 Mton CO_2 equivalent. This fact increased the pressure on the downstream players to improve their policies related to the flaring systems operation in order to minimize the unnecessary gas flare as well as lead some players to carry out capital investment in the installation of systems capable of reducing the flowrate of unnecessary gases to flare.

The processing units of crude oil refineries are designed to operate under stable operating conditions or in a steady state regime; despite this, sometimes these assets face adverse scenarios where operating transients lead to instabilities which demand mass relief aiming to avoid overpressure in the process equipment and process safety accidents. To control the risk of severe process safety accidents related to the overpressure of process equipment above their design limits, it's necessary to rely on a process safety relief system, aiming to allow the quick depressurization of the inventory of the processing units for a safe place in a safe manner.

The flaring systems of a crude oil refinery are designed to receive the discharge of process safety valves and relief systems of the processing units, aiming to ensure the

DOI: 10.1201/9781003472216-6

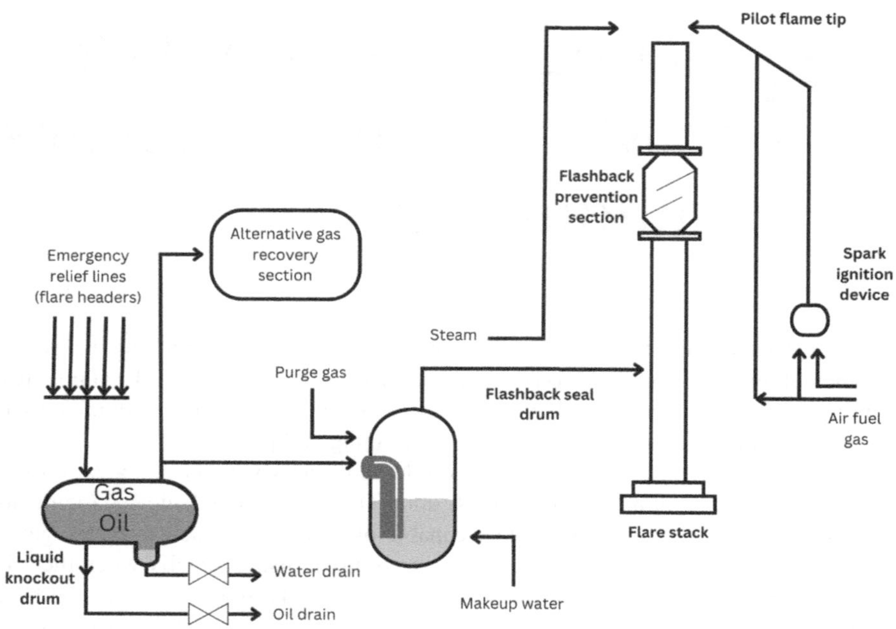

FIGURE 6.1 Typical flaring system of crude oil refineries.

process safety during operating transients like power failure and cooling water failure, which tends to cause overpressure due to mass accumulation in the operating systems. A typical flaring system is presented in Figure 6.1.

The main role of a flaring system is converting the relieved hydrocarbons and harmful gases into less impactful products through combustion in a controlled and safe manner. The flaring system can be classified as described in Figure 6.2.

Most of crude oil refineries apply elevated flares. In this case, the flare tip is elevated from the ground, aiming to reduce the issues related to heat radiation as well as to contribute to the dispersion of burned gases, helping to better control odor emanation and noise issues. The most common flaring system found in crude oil refineries

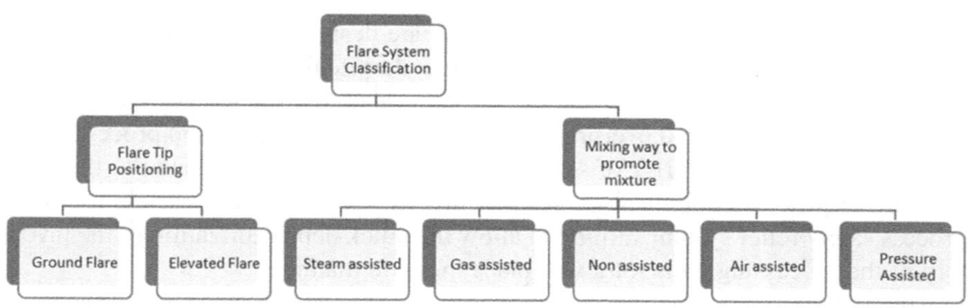

FIGURE 6.2 Classification of refinery flaring systems.

is the single point elevated flare, which presents a low pressure drop. This is a fundamental characteristic and advantage for crude oil refineries that offer low back pressure and consequently help to lower relief systems of the processing units as well as better relief flow.

Despite this advantage, the necessity to achieve a smokeless operation even in emergency scenarios and improve the burning efficiency. The multipoint elevated flare was developed. In this flaring system, the relieved gases are distributed in multiple burners. The main disadvantage of multipoint flares is the relatively high pressure drop in the flare tip when compared to the single point flare.

The ground flaring system applies multiple burners at the ground level, and its advantage in comparison with the elevated flare is easier access for maintenance and less exposure of the flames to the adjacent communities, which can be a significant issue in populated areas. Due to their characteristics, the ground flares present high pressure drops and demand sufficient relief pressure from the process gases for adequate operation and performance. Ground flares normally demand higher capital spending for installation than elevated flares for the equivalent burning capacity.

Regarding the design and operation of flaring systems, one of the main challenges is to reach a smokeless operation. The smoke formation is due to the incomplete combustion of hydrocarbons, leading to the emanation of non-burned carbon, which is black smoke. There is a classification of the intensity of black smoke emanation, which helps the designers to determine how efficient the burning process is. The Ringelmann scale is a measure of the density and opacity of smoke; the scale is summarized below.

- Ringelmann 0–0% opacity
- Ringelmann 1–20% opacity
- Ringelmann 2–40% opacity
- Ringelmann 3–60% opacity
- Ringelmann 4–80% opacity
- Ringelmann 5–100% opacity

Normally, the flaring systems are designed to reach smokeless operation and Ringelmann lower than 1. To achieve this goal, the mixing way to promote the mixture in the flare tip is essential, as normally the smoke or the incomplete combustion is due to a poor mixture between the gases with oxygen.

To overcome a poor mixture of oxygen and gases, it's necessary to promote turbulence and ensure adequate air supply to the flare tip. The way to reach this can be using the following strategies:

- Steam-assisted flares: Steam is supplied to the flare tip, aiming to promote turbulence and drag air to the flame;
- Air-assisted flare: Normally applied in refineries without availability of steam in sufficient quantity;
- Gas-assisted flares: Applied in specific cases, mainly when the availability of steam in the refinery is limited. In this case, a gas (normally natural gas) is supplied at multiple points in the flare, aiming to drag air to the flame and

help to achieve clean and stable combustion. The gas-assisted flare can be an attractive strategy to burn low calorific value gases like ammonia and hydrogen sulfide, as the gas raises the calorific value of the mixture and helps to reach efficient combustion;

- Pressure-assisted flares: In this case, the energy of the relived gases is used to promote turbulence and maximize the mixture with the air, leading to a better combustion process.

Steam-assisted flares are used most commonly in crude oil refineries due to the higher availability, reliability, and low cost of steam when compared to other assisting systems.

Another significant factor which impacts the smokeless operation of the flaring system is the composition of the burning gases. Heavy hydrocarbons, olefins, and aromatic molecules present a higher C/H ratio and tend to produce more smoke during the burning process, demanding a higher flowrate of steam to reach the smokeless operation. It's important to quote those gases produced by deep conversion processes like fluid catalytic cracking (FCC), and delayed coking units tend to present high concentrations of aromatics and olefins in their composition. Based on this, it is expected that an emergency in an FCC and delayed coker will demand more steam flowrate to the flare, aiming to reach smokeless operation than an emergency scenario in a distillation processing unit, for example.

The conventional flares have two points of steam injection: an upper injection and an inferior injection, as presented in Figure 6.3.

The inferior injection points are positioned in the center of the flare tubulation and further assist the combustion process, helping to cool the flare tip, minimizing the flare degradation due to metallurgic damages produced by high temperatures. Metallurgic damage is even more relevant if the flare tip operates burning gases with high concentrations of hydrogen, which can induce stepwise cracking and hydrogen embrittlement.

One of the main operating issues of conventional flares is the poor distribution of the burning mixture in the flare tip for diameters above 36" due to the poor access of the air to the center region of the flare tip, leading to smoke production in this region.

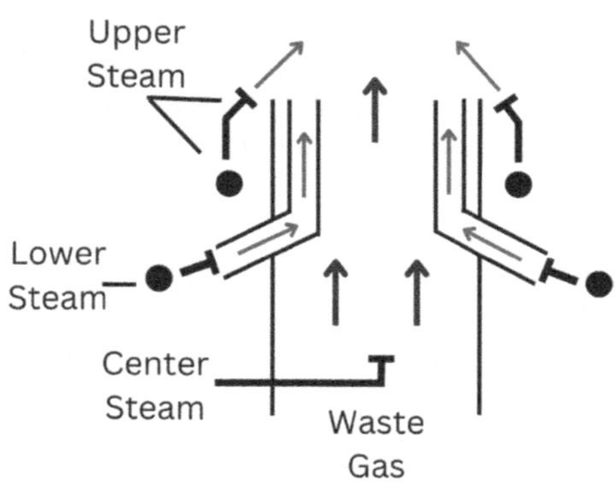

FIGURE 6.3 Steam injection points in a conventional elevated flare.

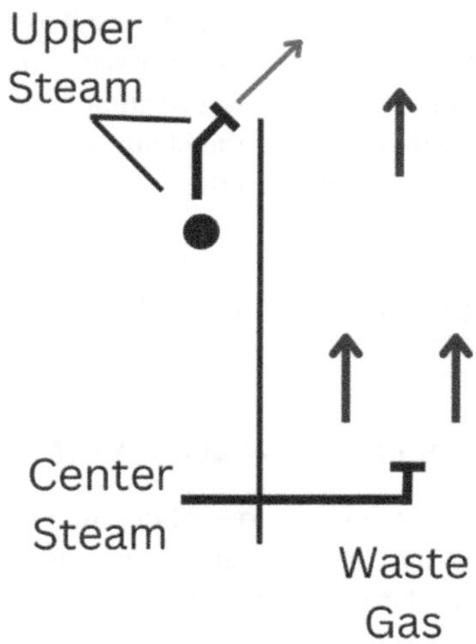

FIGURE 6.4 Steam injection points in a high efficiency elevated flare.

To overcome this, some flaring licensors developed new design concepts, considering new points of steam injection in the inferior region of the flare tip, as presented in Figure 6.4.

The flare tips apply the lower steam injection points as presented in Figure 6.4, which are described as high efficiency flares that once minimize the effect of rich mixture formation in the center of the flare tubulation, reducing the black smoke formation. Among the high efficiency flare technologies available in the market, it's possible to quote the Steam Force HC™ developed by ZEECO Company and the Steamizer XP™ developed by John Zink Hamworthy Combustion Company.

It's important considering that the control of the steam flow rate to the flare tip is fundamental to allow a reliable and efficient operation of the flaring system. An excess of steam in the upper region of the flare tip can lead to the steam capping phenomenon, where the steam excess creates a kind of seal in the flare tip, forcing the flaring gases down and causing a flashback effect. This phenomenon causes overheating of the flare tip and reduces the operating lifecycle of the flare system. In extreme cases, it's possible to occur the flare extinguishing or flame out effect.

The ideal steam flow rate to the flare tip should consider the flame aspect and ensure an adequate balance between the gas velocity in the flare tip and the flame velocity. If the velocity in the flare tip is low, the flashback effect can occur, leading to overheating of the flare tip; if the velocity is high, it's possible that a flame out will occur and the flame will be extinguished with the additional risk of explosion of the flare stack due to the explosive mixture. The flare tip exit speed must be such that the flame forms immediately at the burner exit, aiming to produce an attached stable flame; high velocity can lead to flame out effect and the blow off the flare tip.

Another effect of over steaming is the production of poor mixture and inefficient combustion, which can produce smoke. Furthermore, the over steaming waste energy raises the operating costs. The operating concept of the flare should consider the minimum steam flow rate necessary to ensure the flare tip cooling and protection as well as the additional amount capable of allowing a stable flame without black smoke formation. Furthermore, the over steaming produces excessive aeration of the flame, leading to poor flame quality due to the dilution of the combustion zone. The over steaming process is easily identified through the steam condensation in the base of the flare tip (white plume formation) as well as the transparent and light yellow color of the flame.

On the other hand, the low steam flow rates to the flare tip cause under aeration of the combustion zone, leading to the black smoke formation and the presence of orange-to-red flame.

6.1 MAIN COMPONENTS OF FLARING SYSTEMS

Typical flaring systems are composed of a series of essential process equipment and instruments, aiming to ensure the maximum performance and safety of the refinery. The main components of typical flaring systems are presented as follows.

6.1.1 Stack Sealing System

The main objective of the stack sealing system is to avoid the air entering the flare stack. The air admission can lead to the formation of an explosive mixture inside the flare stack, causing explosions. To prevent this risk, the flare stack operates under positive pressure through the injection of purge gas (normally natural gas), avoiding the entering of oxygen due to wind and thermal contraction effects.

In order to reduce the consumption of purge gas, the modern flaring systems rely on a flashback-preventing system, which can apply two seal technologies. The buoyancy and the velocity seals as presented in Figure 6.5.

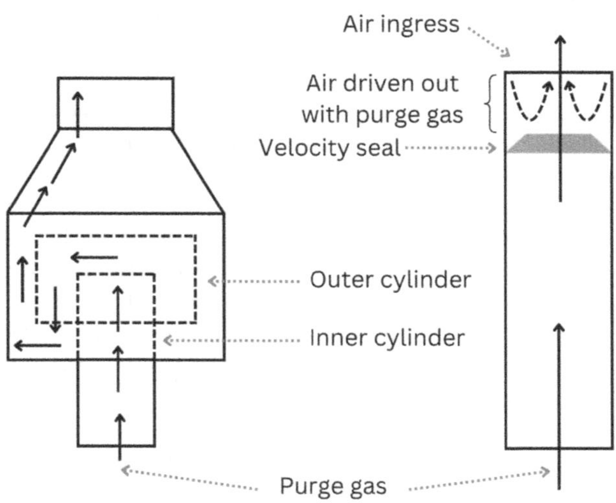

FIGURE 6.5 Typical buoyancy and velocity seals for elevated flares (Based on YOON, B.H. – 2015).

The buoyancy seal principle is based on the difference between the density of air and purge gas. The most common buoyancy seal for elevated flare is an inverted can device which is assembled upstream of the flare tip as presented in Figure 6.5.

The dynamic or velocity seal is sized so that it forces air to move away from the burner walls. This air meets the ascending purge gas and is drawn out of the tip.

The buoyance seal tends to lead to lower purge gas consumption than the velocity seal but presents a higher installation cost and higher pressure drop. Another key question that should be considered in the choice between the buoyance and velocity seal is the higher risk of blocking in the buoyance seals due to the draining line blocking, especially in refineries, which do not have chemical flaring systems.

6.1.2 Ignition System

This system is responsible for the ignition of the flaring gases in the flare tip. The system keeps a group of pilots ignited, which operates continuously. Considering that the pilots are responsible for keeping the flare lit, the gas supply for this system should be as reliable as possible, including supply redundancy as recommended by 521 API. There are two pilot ignition systems commonly applied in refinery flaring systems:

- Flame Front Generator: A flammable mixture of gas and air is produced in a tubulation, which is conducted to an ignition panel and then fed to the pilots.
- Electronic Ignition: The ignition is made in the pilot body through a spark generation.

Considering that this is a critical system to the process safety of the refinery, the flaring system design normally considers both systems as redundant in the flare system. Another key factor in the pilot's ignition system is the pilot's flame detection. The system should be able to distinguish the pilot flame from the flare tip flame, and each pilot needs to have an individual flame detection instrument.

The most common way applied to pilot flame detection is to use thermocouples, but it's possible to use acoustic and radiant flame detectors.

6.1.3 Liquid Separation System

The flaring system is designed to burn gases, and the liquid relieved by the processing units should be separated and recovered to the process, aiming to avoid damage and poor efficiency of the flare tip.

The presence of liquid in the flare can lead to flame extinguishing, flaming rain phenomena (liquid projection in the flare tip), flame instability, black smoke due to rich mixture formation, and higher pressure drop in the flare system.

To remove the liquid from the released gases, the flaring systems rely on a knockout or a blowdown vessel. The vessel is dimensioned to ensure adequate residence time to allow the maximum liquid recovery from the waste gases even due to the vapor condensation from the process gases.

According to the refining scheme of the refinery, some processing units can use internal blowdown vessels, aiming to minimize the presence of the liquid in the flaring system; this is the case of delayed coking and solvent deasphalting units. The process arrangement of a flaring system blowdown vessel is presented in Figure 6.1.

The liquid accumulated in the blowdown vessel is pumped to a residue recovery tank for recovery of the hydrocarbon phase, normally in deep conversion units like FCC, while the water phase is directed to sour water stripping units.

The draining operations to the flaring systems demand specific risk analysis, especially related to the liquefied gas draining operations like LPG and propylene to the flaring header. Some refineries rely on the closed draining system in their liquefied gas storage tank farms (sphere pharms), aiming to remove the water accumulated in the bottom of the spheres. These draining operations should be monitored under a controlled draining rate, aiming to avoid a quick blowdown refrigeration process, which raises the liquefied gas solubility in the liquid, leading to cavitation of the blowdown pumps and causing process safety accidents in case of emergency relief of the processing units. During the draining operations of liquefied gases, it's necessary to monitor the temperature in the blowdown vessel to avoid cavitation of the blowdown pumps.

Some refineries install cyclone separation vessels downstream of the blowdown vessels, aiming to maximize the efficiency of liquid separation from the flaring gases through the condensation of liquid droplets using centrifugal forces. The side effect of the cyclone separation vessel is an additional pressure drop in the flare header.

6.1.4 Liquid Seal Vessel

The liquid seal vessel is positioned between the blowdown vessel and the flare stack. The objective of the liquid seal is to promote a kind of barrier between the released gases and the gases in the stack, aiming to avoid air entering the system and the formation of an explosive mixture inside the flare system. Sudden cooling or gas condensation should create negative pressures, and the role of the liquid vessel is to make the gases bubble inside the liquid, avoiding the flashback occurrence.

The liquid seal is responsible for keeping a positive pressure in the flaring system, avoiding the air entering even in non-emergency operations. It's important to control the liquid seal level to ensure a stable and safe operation, as lower levels can lead to air access to the system, and the explosive formation inside the flare system and high levels can produce high back pressure to the relief system as well as pulsation of the flame. The level of the liquid seal vessel is controlled through the addition or draining of industrial water to the vessel.

6.1.5 Flare Header

The flare header is the tubulation responsible for receiving the relief of the processing units or process facilities and conducting these gases and liquids to the blowdown and flare tip to burn. The design of the flare header should avoid low points capable of accumulating liquid, which can accelerate the corrosion process as well as produce hydraulic transients during emergency discharges of the processing units.

The dimensions and consequently the cost of installation of the flare header can be significantly reduced through actions to reduce the gas flow rate to the flare system in emergency scenarios. Some refineries use water injection over the heat exchangers of the main contributing systems, like the top of the main fractionating tower, like FCC, distillation, and delayed coking units, aiming to reduce the gas flow rate to the flare system and reduce the dimensions of the system while keeping the hydraulic capacity.

6.1.6 Flare Gas Recovery System

The increasing necessity to reduce the carbon footprint of the refining business, as well as the search for more efficient operations, creates the demand by refiners to reduce the gas flaring, aiming to minimize the carbon emissions and reduce operating costs, considering that a great part of the flaring gases can be recovered.

A critical issue to flare systems is the constant relief of waste gas with a high concentration of hydrogen. As aforementioned, the hydrogen can cause severe damage to the flare tip by reducing the operating lifecycle of the equipment. Other reasons to avoid the hydrogen relief to the flare system are economic and environmental impact. The hydrogen generation is an energy intensive process, which produces high amounts of carbon dioxide. For both points of view, hydrogen waste is a serious concern to the modern refining industry. The conventional hydrogen production route adopted by most refiners is methane steam reforming. This route still presents high emissions of greenhouse gases (11 t CO_2/t H_2) and an average production cost of US$ 1.0 to 2.1 per kg H_2.

The demand for hydrogen rose strongly in the last decades following the necessity of hydrotreatment unit installations in refineries to comply with the pressure to reduce the content of contaminants such as sulfur and nitrogen in the petroleum derivates and consequently minimize the environmental impact caused by fuels burned. Hydrogen became a fundamental production input to modern crude oil refineries, and its adequate management is a key factor to ensure controlled operating costs and competitiveness in the market, as well as allow the production of marketable crude oil derivatives. The hydrogen management actions start with a mass balance involving the hydrogen network that is composed of hydrogen sources, hydrogen purification systems, and hydrogen consumers as presented in Figure 6.6.

The hydrogen generation relies on the refining configuration adopted in the refinery. Normally, refineries that rely on catalytic reforming units apply the hydrogen produced in this process unit to compose a relevant part of the hydrogen network, becoming an important internal source of hydrogen.

Hydrogen purifying technologies are another important part of the hydrogen network. Normally, modern refineries apply pressure swing adsorption (PSA) technologies to purify the hydrogen, reaching a purity higher than 99%. Despite this fact, some refiners still use treatments based on amine treatment, although the lower capital cost requirement when compared with PSA technologies, the amine treating units produce hydrogen with low purity, and this represents a great disadvantage, especially to refiners with deep conversion hydroprocessing units. Other hydrogen purifying technologies commercially available are membrane separations, which can

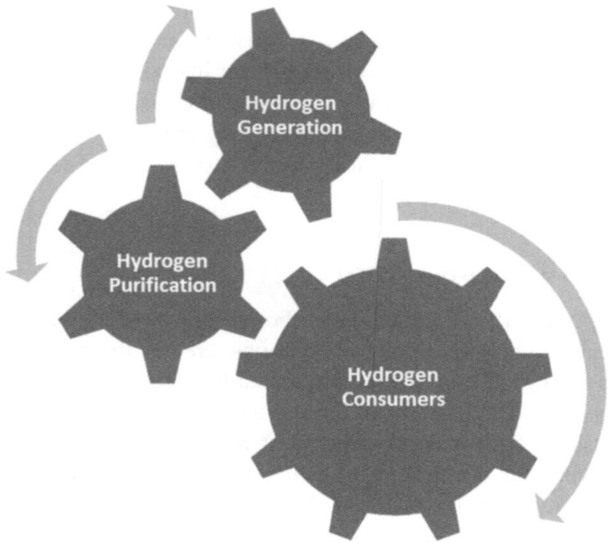

FIGURE 6.6 A typical hydrogen network in a crude oil refinery.

reach a purity of 98%, and cryogenic processes, which can reach 96% purity. Hydrogen purifiers have a key role in hydrogen management. Once control of the hydrogen recovery in off-gases is established, one of the main sources of hydrogen losses in the refineries is the burning of fuel gas during poor recovery capacity.

The adequate balance between the production and consumption of hydrogen is a key factor in minimizing the hydrogen flaring. It's possible to apply optimization techniques like the pinch strategy to reach an adequate balance between producers and consumers of hydrogen in the refinery, as presented in Figure 6.7.

The flaring gas recovery system applies a compressor to suction the gas phase of the blowdown vessel and sends it to the fuel gas system of the refinery or, for hydrogen-rich gas, to a PSA capable of promoting the hydrogen recovery to reuse in the refining processes. A typical flare gas recovery system is presented in Figure 6.8.

As described above, the refiners are under pressure to reduce carbon emissions, and the competitive environment of the downstream industry demands increasing operational efficiency from the downstream players. Under this scenario, the flare gas recovery system tends to be essential to the refinery operation in the short term.

The greater issue related to the operation of the flare recovery gas system is the poor control and reliability of the compressor due to the constant variations in flow rate and composition of the gas. Among the strategies available to overcome this issue is to apply liquid-ring compressors to this service or even educators using amines as drive fluid.

Another strategy in alternative to a central flare recovery gas unit is using lower capacity dedicated systems installed in the battery limits of each contributor. This strategy will allow better control of the gas composition and better operating performance for the flare gas recovery compressors.

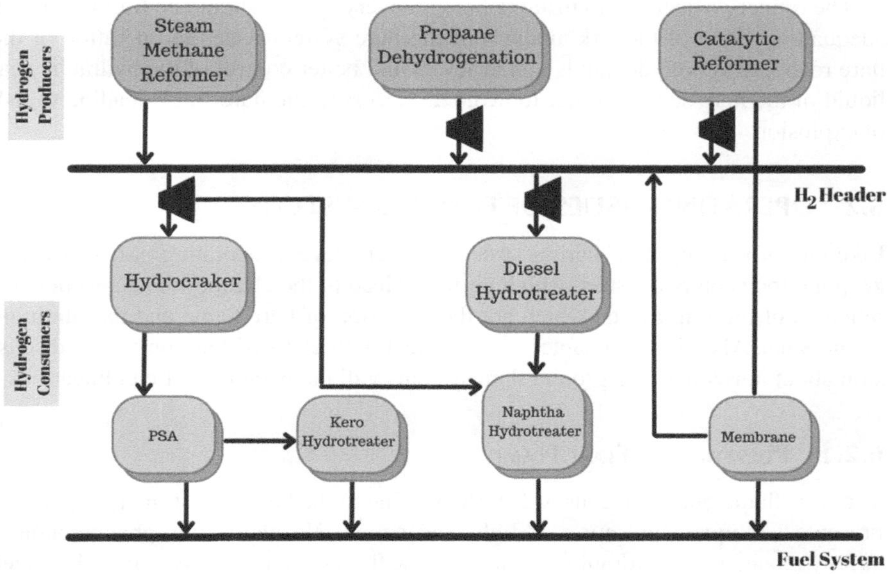

FIGURE 6.7 Example of hydrogen network in a crude oil refinery.

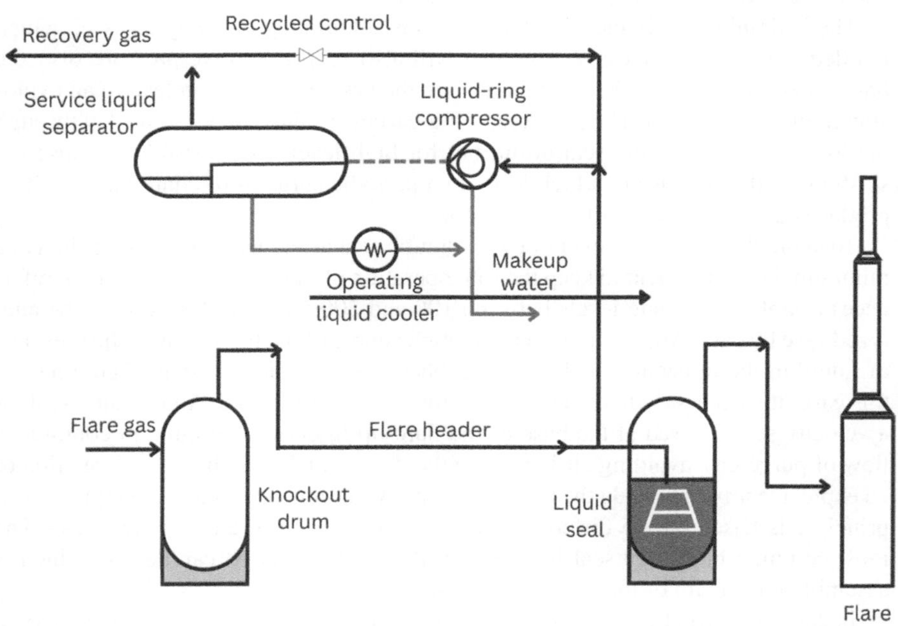

FIGURE 6.8 Typical flare gas recovery system.

The refinery's decision to install a flare recovery system should be followed by an adequate revision of the risk analysis of the flare system once the operation of the flare recovery system demands higher levels and better control of the hydraulic seal liquid of the flare stack in order to avoid air access to the flare stack, leading to risk of explosion.

6.2 OPERATING ISSUES OF FLARING SYSTEMS

Like any processing unit, flaring systems present several operating issues. We will keep the focus on two issues, which normally lead to the claim of the adjacent communities of refining assets, which are the pulsation of flare flame and the rumbling phenomena. Also, in this chapter, we will deal with elevated flare only. The discussion about horizontal and ground flare systems will be made in future publications.

6.2.1 Pulsation of Flare Flame

The flare flame pulsation causes disturbance due to the behavior of the flame, which presents a frequent pulsation of light and sound. Normally, this phenomenon is strictly related to the hydraulic seal level of the flare system. This system is dedicated to avoiding a flame return (flashback) from the flare burn; this is especially a concern when the system is operating under low gas flow rates, which lead to the low velocity of the mixture that can be lower than the flame velocity. This will lead to a burn inside the burner, which can lead to an explosion.

The hydraulic seal in elevated flare systems is normally made by a vessel, which is filled with a level of water. This seal will also keep a positive pressure over the flaring system, avoiding the entry of air into the system, which can lead to an explosion in the flare system. The pressure of the flaring system can be controlled through the level of the hydraulic seal drum, but for high levels, it's possible to cause the sloshing of the seal drum, which leads to a periodic surge in the flare flame, which produces noise and instability of the flame.

To avoid this issue, the level of the hydraulic seal drum needs to be controlled at a minimum level sufficient to keep the safe operation of the flaring system. Some references establish adequate levels between 100 and 200 mm, but this needs to be analyzed case by case. Another possible vibration source of the flame can be the presence of liquid in the molecular seal of the flashback prevention section in Figure 5.6. In this case, it's necessary to check if the drain line is not blocked. A buoyancy seal, or a velocity seal, is used at the base of the flare tip to ensure a minimum continuous flow of purge gas, avoiding air ingress to the flare, which can allow the formation of an explosive mixture inside the flaring system. As aforementioned, the buoyancy seal principle is based on the difference between the density of air and purge gas. The most common buoyancy seal for elevated flare is an inverted can device, which is assembled upstream of the flare tip as presented in Figure 6.5.

In case of obstruction of the drain line of the molecular seal, it can be necessary to shut down the flare system to clean the drain line; this is especially important to refineries that operate without chemical flaring systems that are dedicated to burning acid and ammoniacal gases from sour water stripping and/or sulfur recovery units

(SRU). The burning of acid and ammoniacal gases in a conventional flaring system can accelerate the salt deposition in the draining system.

Another factor which can lead to flame vibrations in flaring systems is an excess of steam flow rate injection. Steam injections are used to reduce the smoke formation during the flare operations, especially when heavy hydrocarbons are released, and it is normal for the refiners to raise the steam flow rate until it reaches a transparent flame. Further, in addition to being a waste of energy and money, the excess steam causes pulsation, noise, and instability of the flare flame. To avoid this, the steam flow rate should be reduced to the minimum necessary to produce a yellow flame without smoke.

Another design strategy which can be applied to control and minimize the flame pulsation in the flare is the use of anti-slosh baffles inside the liquid seal vessel. In some cases, mainly for high dimensions of liquid seal vessels, there is a sloshing effect during the gas flow through the liquid seal, which causes an oscillatory movement of the liquid and then flame pulsation.

6.2.2 Rumbling Phenomena in Elevated Flare Systems

The rumbling phenomenon occurs when excess air is fed to the flare tip. The over aeration of the flame causes their lift from the flare tip, followed by reattachment of the flame as the pilots relight the flame. This periodic and quick lift and reattachment of the flame leads to a low-frequency vibration or rumble, which can produce a shock wave propagation that can produce vibration of adjacent structures like windows and doors; this is a special concern for crude oil refineries that are installed in areas with high population density.

The rumble phenomenon can be avoided through the adequate control of steam flow rate to the flare tip; as discussed above, the steam is applied to minimize the smoke formation during the flaring operation. In some flare designs, ejectors are installed in the flare tip in order to drag air to the flame, promoting air excess to the flame and then reducing the smoke formation. This design is normally applied in refineries with the limitation of steam availability.

6.3 CHEMICAL FLARING SYSTEMS

Some refineries apply separated flaring systems to deal with hydrocarbons and acid or ammoniacal gases. The chemical flaring system is designed to deal with process gases with high concentrations of H_2S and NH_3. This system is used to receive the relief of treating units like sour water stripping and SRUs.

Sour water is the water that had contact with the petroleum or its derivates during some step in the process. This contact can be like rectification steam in distillation columns or in contact with hydrocarbon phases. Contaminants like NH_3 and H_2S tend to concentrate in the aqueous phase, so sour water commonly has a high concentration of these compounds.

The sour water stripping unit applies the concept of fluid rectification with steam and the partial pressure reduction to move the phase equilibrium to the vapor phase, releasing the contaminants from the liquid, as presented in Figure 6.9.

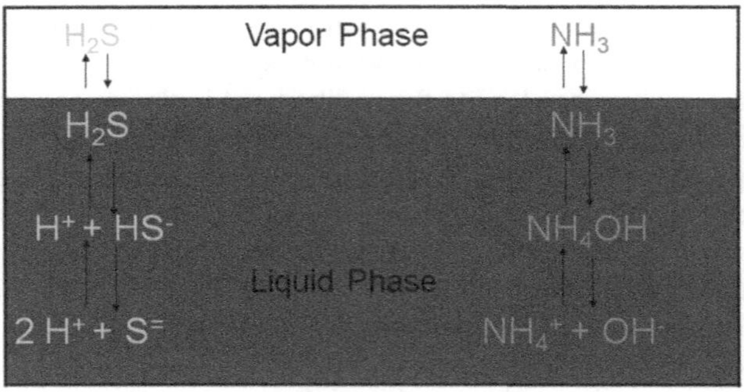

FIGURE 6.9 Phase equilibrium in sour water stripping process.

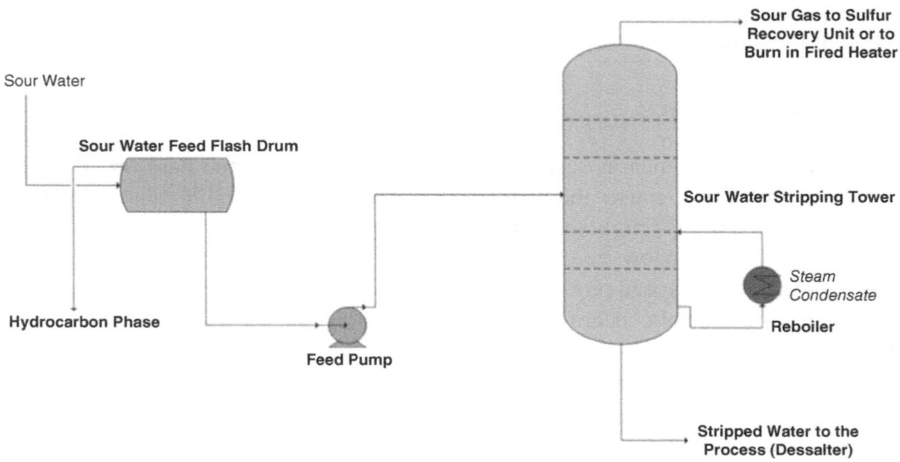

FIGURE 6.10 Typical arrangement for sour water stripping unit with single tower.

Like any other process or technology, the sour water stripping units were developed and improved over time, mainly to reduce atmospheric emissions and to raise water reuse in the refineries.

The initial design concept for sour water stripping units had one rectifying tower. In this tower, both contaminants (NH_3 and H_2S) were removed and formed the stream called sour gas, as described in Figure 6.10.

In these cases, the tower operates with relatively low pressure (about 1.0 kgf/cm^2).

Initially, the designs predict sending the sour gas to burn in fired heaters, such as in distillation units. Nowadays, with environmental restrictions and the necessity to reduce SOx and NOx emissions, the project concept has changed, and the sour gases are directed to SRUs with a chamber to convert the NH_3 to N_2. This is necessary to avoid that the NH_3 prejudices the H_2S conversion in elemental sulfur through the Claus process.

The modern designs rely upon the installation of two towers, one for the H_2S removal and the second for the NH_3 removal, as described in Figure 6.11.

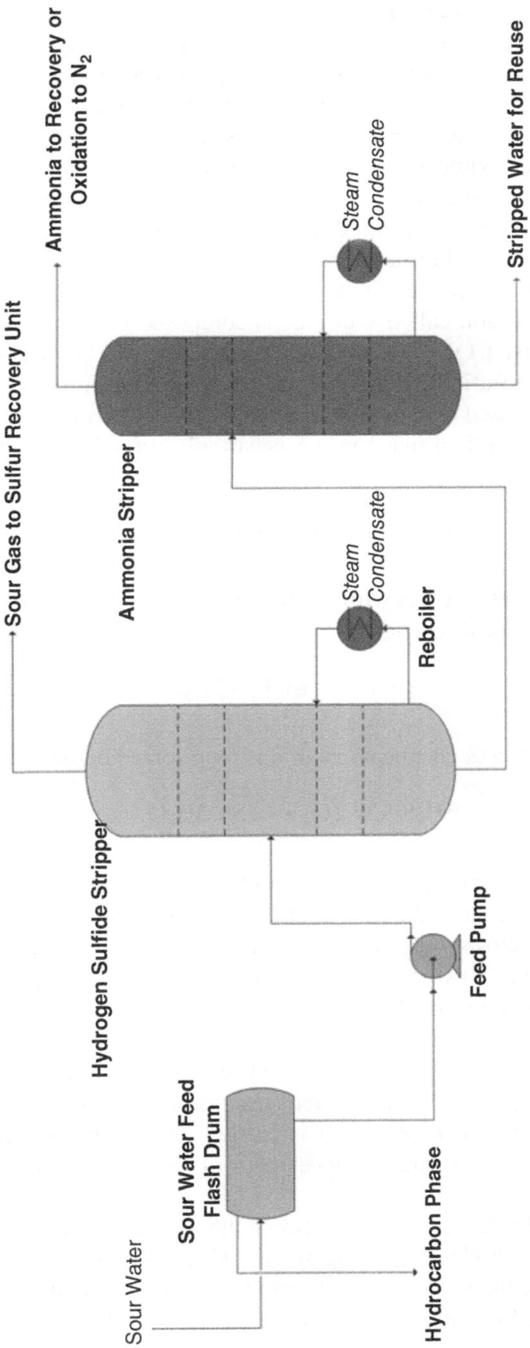

FIGURE 6.11 Typical arrangement for sour water stripping unit with two towers.

For units with two towers, the H_2S rectifier operates under pressures of about 5 to 11 kgf/cm^2, while the ammonia rectifier operates under pressures of about 1 to 2 kgf/cm^2. The arrangement with two towers shows some advantages in relation to the project with a single tower, since that allows higher recovery of H_2S-like elemental sulfur, reducing the SOx emissions. Furthermore, the design with two towers allows the ammonia present in the sour water to recover or convert this stream to N_2.

The SRU is responsible for recovery. In the elemental sulfur form, the sulfur removed from process streams treated in sweetening units of light fractions (LPG and fuel gas) was produced in the deep conversion units such as delayed coking, FCC, and hydrotreating. Furthermore, the gaseous streams produced in the sour water stripping unit are directed to the SRU.

The SRU feed stream, called sour gas, is composed basically of H_2S (50–80%) and contaminants like CO_2, H_2O, NH_3, and hydrocarbons. The most employed technology to recover sulfur in the refining industry is the Claus process.

Claus process is based on two H_2S conversion steps: a thermal step, followed by a catalytic step. In the thermal step, the H_2S is partially burned according to the chemical reaction below:

$$H_2S + 3/2\,O_2 \rightarrow SO_2 + H_2O \tag{6.1}$$

Then, the remaining H_2S reacts with the SO_2, producing elemental sulfur according to the following chemical reaction:

$$2H_2S + SO_2 \leftrightarrow 3S + 2H_2O \tag{6.2}$$

The global Claus process chemical reaction is represented below:

$$3H_2S + 3/2\,O_2 \leftrightarrow 3S + 3H_2O \tag{6.3}$$

Figure 6.12 shows a process flow diagram for a typical sulfur recovery unit.

The thermal step is carried out in the burner, which operates under temperatures higher than 900°C; close to 1/3 of H_2S is converted to SO_2, following reaction 1, which is endothermic. This step is also responsible for destroying sour gas contaminants such as ammonia and hydrocarbons. The thermal step is responsible for 60–70% of the sulfur recovery.

The catalytic step is realized in fixed bed reactors containing TiO_2 or activated alumina as the catalyst. The catalytic step (reaction 2) is slightly endothermic when compared with the thermal step, so it is carried out at lower temperatures (200–350°C). One of the main catalyst developers of the Claus process is the Axens Company.

The cooling of the process stream between the thermal and catalytic steps is realized in a waste heat boiler producing steam, which is sent to consumers in other processes in the refinery. The process applies multiple reaction stages with the removal of produced sulfur among the stages, aiming to shift the chemical equilibrium to the products. In units containing three catalytic stages, it is possible to recover 98% of the sulfur contained in the sour gas that is fed to the unit.

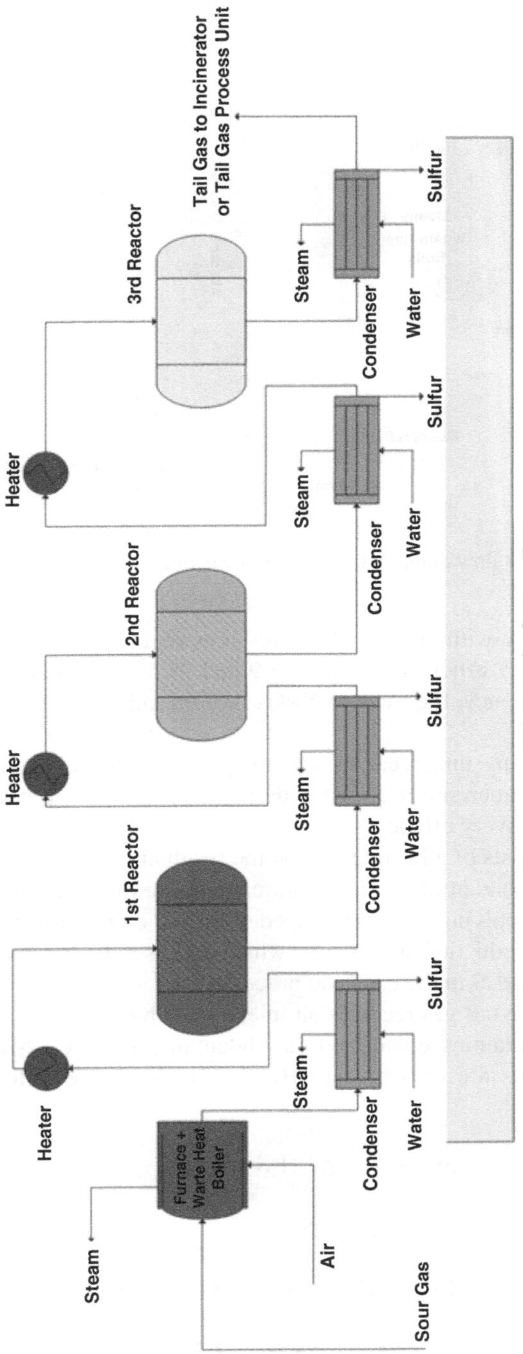

FIGURE 6.12 Process flow diagram for a typical sulfur recovery unit.

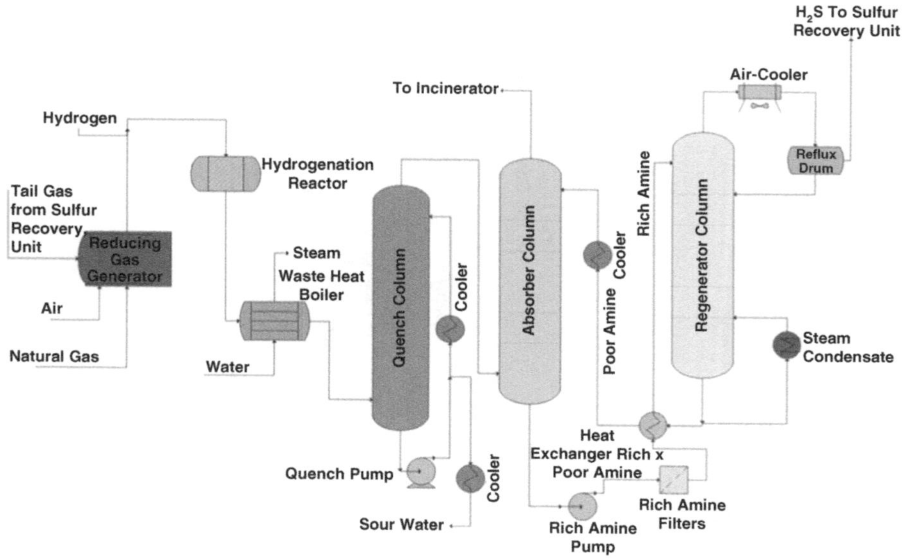

FIGURE 6.13 Process flow diagram to typical tail gas treating unit.

In order to comply with current SOx emissions regulations, the SRUs normally need a sulfur recovery efficiency between 99 and 99.5%. With an aim to raise the sulfur recovery efficiency, the modern SRUs rely on tail gas treating units, as presented in Figure 6.13.

The tail gas treating unit receives the off gas from the SRU and converts the remaining SO_2 and other sulfur compounds in H_2S that are sent back to the SRU, raising the sulfur recovery efficiency.

This process consists of a heating step of the residual gas that raises the temperature over the sulfur condensation temperature, avoiding this phenomenon occurring in the reactor and supplying the energy needed for the conversion reactions. In this step, the hydrogen production still occurs, which acts as reduction gas to convert the sulfur compounds to H_2S in the catalytic process step.

Following this, the tail gas receives an injection of hydrogen before entering the fixed bed reactor containing cobalt and molybdenum catalyst (CoMo). In this step, the sulfur compounds are converted to H_2S according to the following chemical reactions:

$$SO_2 + 3H_2 \rightleftarrows H_2S + 2H_2O$$

$$S + H_2 \rightleftarrows H_2S$$

$$COS + 4H_2 \rightleftarrows CH_4 + H_2S + H_2O$$

$$CS_2 + 4H_2 \rightleftarrows CH_4 + 2H_2S$$

H_2S produced is processed in amine-treating columns to purify this compound, and then the H_2S is sent back to the SRU, and other gases are sent to an incinerator.

Considering the high concentration of NH_3 and H_2S in the relived gases from treating units like those described above, the chemical flaring systems are designed to burn low flammability gases, demanding special flare tips with high flame temperature.

Another characteristic of the chemical flare systems is that the discharge header should be heated and isolated, aiming to minimize the risk of salt deposition and consequently corrosion due to NH_4HS caused by the chemical reaction between the H_2S and NH_3.

The use of chemical flares is highly recommended, especially in densely populated regions, once the conventional flares do not have adequate temperature and design to destroy the NH_3 and H_2S, and it's possible to face issues with odor and claim from adjacent communities. Furthermore, the salt deposition in liquid and buoyance seals can block the flare stack, leading to a reduction in the operating lifecycle of the flaring system.

BIBLIOGRAPHY

American Petroleum Institute, API Standard 537, *Flare Details for General Refinery and Petrochemical Service*, API Standard 537, Draft 3rd Ed., Sept. 2001.

American Petroleum Institute, *Guide for Pressure-relieving and Depressing Systems*, API Recommended Practice 521, 6th Ed., January 2014.

Energy Transitions Commission (ETC) – *Making the Hydrogen Economy Possible: Accelerating Clean Hydrogen in an Electrified Economy*, 2021.

Fahim, M.A.; Al-Sahhaf, T.A.; Elkilani, A.S. *Fundamentals of Petroleum Refining*. 1st ed. Elsevier Press, 2010.

John Zink Co., *The John Zink Hamworthy Combustion Handbook*, Vol. 3, Applications, Chapter 11: Flares, CRC Press LLC, 2001.

Myers, R.A. *Handbook of Petroleum Refining Processes*. 3rd ed. McGraw-Hill, 2004.

Yoon, B. H. *Optimizing Flare Operation Through Proper Design*, Chemical Engineering Magazine, October 2015.

7 Process Safety Management in Transfer and Stockpiling Operations

The stockpiling assets deal with a great amount of energy, which is stored in tanks and pipes; due to this reason, the most adverse scenarios related to process safety can occur in the transfer and stockpiling area.

Unfortunately, due to operating normally under low pressure and temperature, there is a trend of underestimating the safety risks in stockpiling operations by operators, managers, and engineers, and this needs to be adequately managed, aiming to minimize the risks of accidents with great impact, like fire in hydrocarbon storage tanks or contention loss in pipelines.

Fortunately, process safety management in the crude oil refining industry presented great advances in the last decades, but the cost of this development was the loss of human life in process safety accidents that have marked the downstream industry, like the Texas City refinery in 2005 and the Formosa plastics petrochemical plant in 2004. A series of process safety accidents, even without causing loss of human life, caused serious damage to the installations, and in some cases, the process plant was definitively closed, impacting several lives and local economies, like the case of the Philadelphia Energy Solutions (PES) refinery in 2019.

An adequate process safety management system is responsible for keeping under control the inherent risks associated with the petroleum refining process. The success of any process safety management system depends on all the elements of the productive chain from the material suppliers, engineers, operators, managers, etc. Once each one is responsible, or your activities are related to at least one of the process safety barriers.

Throughout the years, the process industries advanced in their approach related to process safety from a reactive approach based on rules and standards to a risk-based strategy.

The standard-based approach was developed considering the lessons learned from process safety accidents to develop safer procedures, standards, and rules, aiming to minimize the risk of recurrence of process safety incidents. The evolution to risk-based strategies allows the organization to follow the regulatory requirements and keep learning from past accidents, but the approach allows the organization to anticipate management actions to avoid process safety accidents, considering information which describes that point of the process that is under higher risk.

DOI: 10.1201/9781003472216-7

FIGURE 7.1 Pillars of risk-based process safety.

With this information, the organization can take actions capable of eliminating or putting under control the process safety risk before an incident or accident. According to the CCPS (Center for Chemical Process Safety), the risk-based process safety is based on four pillars, as presented in Figure 7.1.

Still, according to CCPS, the process safety commitment of an organization is translated into five elements:

- Process safety culture.
- Compliance with regulatory requirements.
- Process safety competency.
- Workforce engagement.
- Stakeholders outreach.

A process safety culture can be resumed as a synergy between behaviors and values, which determines how process safety will be managed in the organization (CCPS, 2007). An adequate process safety culture will lead the workforce and managers to be intolerant of deviations and risks of process safety, preventing the organization from considering coexistence with process safety deviations and risks.

Compliance with regulatory requirements is a fundamental part of the process safety commitment by the organizations, and this is even more important to the downstream industry, which is highly regulated due to the inherent risks associated with crude oil refining and petrochemical production. The refiners need to keep up to date with regulations through continuous capital investments in operational improvements, aiming to keep the refining assets in compliance with the last revision of standards like API and NFPA requirements.

The development of process safety competency is related to improving continuously the knowledge of process safety, ensuring that relevant information will be available to the stakeholders, and applying in an appropriate manner the lessons learned with the experience during the development of process safety competency. This is strictly related to training and knowledge, especially the operating and maintenance personnel which are directly involved with the process safety risks.

Workforce engagement is another essential part of the culture of process safety in the organization. The engagement process involves the active participation of the whole workforce in the development of a process safety culture and not only the compliance with rules imposed by the management team, but this is also translated by the search for an interdependent behavior, as described by the Bradley Curve™ proposed by Dupont Company in 1995. In this stage, the organization faces the process of safety as a collective responsibility, and the rules are accomplished because the persons trust in the process, not by obligation.

The stakeholder's outreach is related to the transparency of the organization in relation to the inherent risks of their operations as well as emergency scenarios and adequate communication to these stakeholders.

The second pillar of the risk-based process safety is understanding the risks and hazards of the operations. This pillar is based on two critical elements:

- Hazard identification and risk analysis: The conscience of hazards and risks based on adequate analysis allows the organization to prioritize projects, investments, inspections, and maintenance, aiming to minimize the most severe risks and hazards.
- Process knowledge management: It's fundamental that the employee knows the inherent risks of the process as well as the emergency scenarios and how to mitigate these risks.

The risk management pillar is sustained by nine elements, which are dedicated to ensuring stable and safe operations, aiming to control the inherent risks of the operation (e.g., respecting the operation limits and restrictions), managing adequately the changes in order to keep the risks under control, and offering adequate response in case of accidents or incidents. The nine elements of the risk management of risk-based process safety are presented below:

- Operating procedures.
- Safe work practices.
- Asset integrity and reliability.
- Contractor management.
- Training and performance assurance.
- Management of change.
- Operational readiness.
- Conduct of operations.
- Emergency management.

The last pillar of the risk-based process safety is learning from the experience; this pillar involves actions that ensure that the organization is under constant learning and development in process safety with their own occurrences or from other companies' accidents. The elements which sustain the learning from experience pillar are:

- Incident investigation: Each occurrence needs to be deeply and adequately analyzed, aiming to determine the root causes and produce effective knowledge and actions that will avoid similar events in the future.
- Measurement and metrics: Develop and monitor adequate Key Performance Indicator (KPIs), aiming to allow an effective follow-up of the process safety deviations.
- Auditing: Periodic verification of how effective the process safety management of the organization is.
- Management review and continuous improvement: Closing the PDCA cycle with the verification of the process's effectiveness and proposing revisions for improvements.

One of the key factors of process safety in crude oil refineries is how the operation team knows process safety management, the main process safety risks, and the key safety elements. After the identification of the key safety elements, it's necessary to define the mitigation and prevention barriers capable of preventing the risk scenario or process safety accident. This can be made using one of the traditional hazard identification techniques described below:

1. Semi-quantitative techniques: In this category, the most common technique is the layer of protection analysis (LOPA). This technique helps to identify the frequency of events occurrence, which can cause a determined hazard, the probability of failure of the protection layers, and the consequences, leading to an estimation of the risk associated with the scenario studied.
2. Quantitative techniques: The most applied quantitative technique in crude oil refineries is quantitative risk analysis (QRA), which involves data collection and analysis to estimate the risks of hazardous events, helping to understand the outcomes of different scenarios and then facilitating the decision-making process.
3. Qualitative techniques: In this category of risk studies, there is a series of techniques capable of helping the refinery management to better understand the process safety barriers, which are:
   - Hazard and Operability Study (HAZOP): This is the most common risk assessment technique applied in the refining industry. The HAZOP technique is carried out as a structured analysis of existing or future processes in order to identify and evaluate the associated risks which can affect the people, process, and installation, proposing actions to mitigate the risks.
   - Hazard Identification Studies (HAZID): This technique is applied systematically, aiming to identify potential hazards and their consequences over the process asset, which can involve hazards nonrelated to the process.
   - Control Hazardous Studies (CHAZOP): This technique is a systematic risk assessment approach focused on identifying hazards and operability issues associated with a control system.

- Simultaneous Operation Studies (SIMOPS): This technique is dedicated to identifying hazards and operability issues in simultaneous operations. This technique is especially important to identify process safety risks in stockpiling and transfer operations, which normally involve at least two different operating areas.

The hazards identified by SIMOP studies are normally not considered by the other risk assessment techniques which are dedicated to analyzing individual process operation scenarios. Considering the characteristics of stockpiling operations, the SIMOP study is fundamental to help identify hazards related to interdependent actions.

A very good example of the applicability of SIMOP studies is the operation of filling a storage tank, which is one of the most common operations in stockpiling areas. There are several risks associated with this operation, such as overfill, overpressure, or the reception of a contaminated stream in the tank, which can lead to process safety risks like the mixture of naphtha and LPG in an atmospheric tank due to the operational instability in a processing unit like distillation or FCC. The SIMOP study is a fundamental tool to identify hazards in the transfer and stockpiling operations of a crude oil refinery, mainly related to the lack of integration between the operating teams involved in the operation.

The main steps of a typical SIMOP study are summarized below:

- Identify activities: In this step, the activities or tasks needed to reach a specific objective, like filling a storage tank, are identified and listed. All activities needed in all areas involved should be identified, even those related to the adjacent facilities.
- Collect information about the activities: This step aims to collect the maximum information in order to identify potential conflicts between the operating steps and associated hazards.
- Identify possible interactions: This step is dedicated to identifying adverse interactions between the activities, like process connections, disabled safeguards, and communications failure.
- Identify possible consequences: This step aims to identify the potential impacts of the adverse interactions.
- Identify existing safeguards: A deep analysis is carried out, aiming to identify the existing safeguards to the impacts of the adverse interactions identified in the previous steps as well as identify the required new safeguards which need to be implemented to avoid the identified hazards.
- Identify recommendations for any necessary risk controls: Made recommendations to control the identified risks considering the hierarchy of hazards.

Based on the concepts of risk-based process safety (RBPS), it's possible to propose an approach to the process safety management in stockpiling operations based on three main pillars: asset management, discipline, and integration between the stockpiling area and another operational area, as presented in Figure 7.2.

FIGURE 7.2 Pillars of process safety management in stockpiling operations.

7.1 ASSET MANAGEMENT

Unfortunately, the obvious needs to be said! It's impossible to ensure process safety without an adequate asset management strategy, mainly in transfer and stockpiling operations. As previously mentioned, asset management is related to the pillar of risk management of the risk-based process safety methodology proposed by the CCPS. The asset management policy adopted by a refinery reveals the maturity of the process safety culture and the commitment to this refinery with the process safety.

The contention loss in simple equipment like a tank mixer can lead to a severe process safety accident. Due to the characteristics of the stockpiling area, an inadequate asset management policy will not cause an immediate impact on production, and this fact can lead the maintenance team to reduce the sense of urgency to deal with the reliability and maintenance issues of stockpiling assets.

The stockpiling manager should be able to use their knowledge of the refinery process chain to call attention to the stockpiling assets, aiming to avoid asset degradation.

According to the literature (CHANG and LIN, 2006), a great part of the accidents that occurred in storage tanks between 1960 and 2003 were caused by maintenance errors or equipment failure; the main cause of accidents in storage tanks was caused by lightning. This is interesting data once the lightning accidents can be related to ground system failure, which is also related to asset management issues.

Failures in asset management policy can be the cause of process safety accidents that are initially attributed to human error. Still, according to CHANG and LIN

(2006), most process safety accidents that occurred between 1960 and 2003 and were attributed to operational error are related to tank overfilling, which again can be related to asset management issues, especially related to tank gauging system failures. The stockpiling operating team should be trained to understand that without adequate assets, the operations can't be carried out; for example, a transfer operation should not be started if the tank gauging system is unavailable and should be stopped if the gauging system fails during the transfer operation.

Asset management in stockpiling assets is a special challenge in a crude oil refinery due to the number of equipment and the dimensions involved in this system. We dedicate a chapter exclusively to this topic in this book, but we understand that it's impossible to talk about process safety management in stockpiling operations without an adequate asset management policy.

Stockpiling managers should be able to attract maintenance resources to achieve the best availability and reliability of the stockpiling assets, aiming to offer the refinery the maximum operational flexibility, which will lead to maximum profitability. Unfortunately, this is not an easy task; the common mind map of the maintenance managers is to deserve less attention to stockpiling assets than to processing unit assets. One of the main roles of stockpiling managers is to find creative ways to force the maintenance team to understand the real impact of stockpiling assets over the results of the refinery. To achieve this, it's fundamental to develop in the operation team a culture of intolerance with the operation of poor reliability equipment or even broken equipment and instruments.

The operating team needs to help the refinery to develop the asset management policy in order to avoid sudden failures that raise the safety risks. The adequate evolutionary path of asset management is shown in Figure 7.3.

In Figure 7.3, accidental maintenance is characterized by reactive maintenance, where the approach is just to fix failures. In this approach, there are higher process safety risks once sudden failures occur, which normally depend on the operator's skills to avoid impacts on the process.

Scheduled maintenance is an evolution of accidental maintenance once there is an attempt to anticipate the failure, reducing the dependence of operators skills to avoid

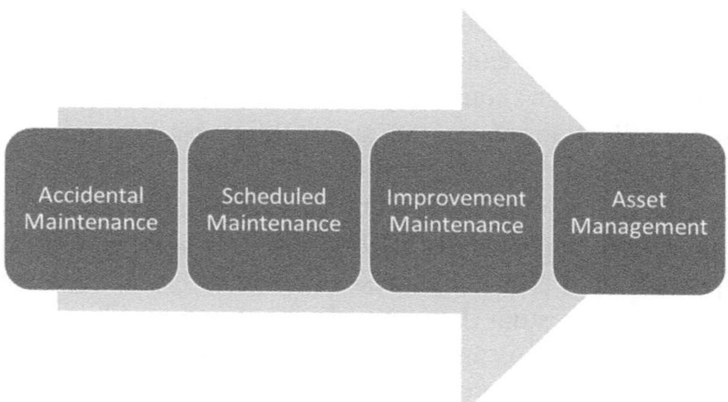

FIGURE 7.3 Evolutionary path of asset management.

process safety risks and the impact over the production. The improvement maintenance aims to eliminate failure through the action in the process, especially those with chronic failures.

The final of the evolutionary process of maintenance in a crude oil refinery is the real asset management policy. In this step, the resources and management actions aim to keep and extend the lifecycle of operating assets as well as improve the asset performance, leading to lower risks of sudden failures and process safety risks.

To improve asset management performance, the stockpiling managers need to sponsor and organize reliability meetings to discuss the chronic failures in equipment and systems, especially those considered critical to the process safety, like the tank gauging systems.

The protection layers of the tank farm should be completely available and reliable in order to minimize process safety risks. As described above, a great part of the process of safety accidents involving storage tanks is related to overfill occurrence. This data reinforces the relevance of these systems to process safety requirements of tank farms once demand continuous monitoring with adequate accuracy and reliability.

According to the API 2350 (Overfill Prevention for Storage Tanks in Petroleum Facilities), the storage tanks of a crude oil refinery can be classified according to their installed level monitoring system; this classification is summarized below:

- Category 0: In this case, the overfill prevention system (OPS) is totally manual, and there aren't instrumentation systems capable of transmitting alarms or liquid level information. In this case, the overfill control is totally dependent on the operator. The API allows these facilities to operate under continuous monitoring during the first hour of the receipt, every hour during the receipt, and again continuously during the last hour of the receipt. Despite this, this author simply understands that it is not acceptable to operate hydrocarbon storage tanks in this way.
- Category 1: This category is related to storage tanks that have only local liquid level information without information transmission to the panel operator.
- Category 2: This category involves the storage tanks which have automatic gauging systems (ATGs) with High–High-level alarms (LAHH) based on the ATG information. The level and alarm information are continuously transmitted to the panel operator. This system is commonly found in crude oil refineries, but there is a fragility once the High–High-level alarm depends on the ATG information. In other words, there is no redundancy in the level control.
- Category 3: In this case, the storage tank relies on automatic tank gauging (ATG) with an independent High–High-level alarm with information transmission to the panel operator. This category is safer once it eliminates the dependence on the High–High-level alarm from the ATG, establishing a redundancy in the level control.

As presented above, each facility category presents an overfill risk level, which demands different response times. The response time is defined as the demanded

TABLE 7.1
Minimum Response Time

Tank Category	Minimum Response Time
0	60 minutes
1	45 minutes
2	30 minutes
3	15 minutes

time necessary to end a receipt in the safest way. This time is calculated based on the tank levels of concern (LOCs), as defined according to the API 2350 criteria. Table 7.1 presents the typical response times according to the storage tanks category.

Normally, the overfill systems applied in crude oil refinery storage tanks should present at least a safety integrity level (SIL) at 2, in other words, a frequency of dangerous failures per hour between 10^{-7} and 10^{-6}. This can be applied using a redundant system allying a level switch and a radar gauge integrated by a PLC. Figure 7.4 presents an example of a safety interlock structure applied in tank farms of crude oil refineries (Category 3 facilities).

The ESD PLC is dedicated to acting in emergency scenarios, applying at least SIL 2 gauging devices, and the radar gauges should present a minimum accuracy of ± 1.0 mm. The strategy proposed in Figure 7.4 presents redundant level control, aiming to avoid overfill accidents which can lead to great emergency scenarios like what occurred in the Buncefield terminal accident in 2005 (United Kingdom) that led to the destruction of the asset.

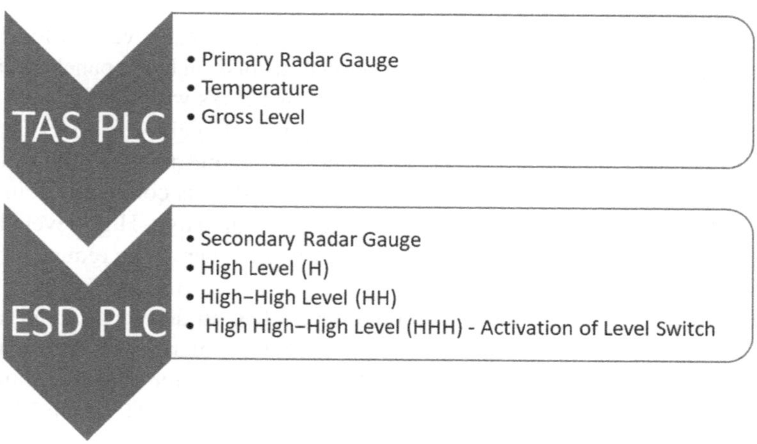

FIGURE 7.4 Typical interlock structure of overfill system of tank farms of crude oil refineries, where: TAS – terminal automation system; ESD – emergency shutdown; PLC – programmable logical computer.

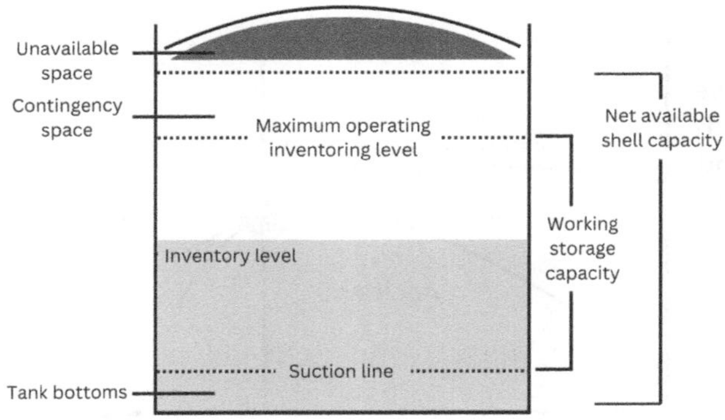

FIGURE 7.5 Tank level control sets focused to avoid overfill occurrences.

The operation team should be as critical as possible with the performance of the stockpiling assets, never accepting operating pumps under vibration or leaking, tanks without gauging systems, grounding, or systems under hydrocarbon leakage, for example. The development of adequate KPIs related to the availability of stockpiling assets should be encouraged and supported by the refinery management team, aiming to achieve a sustainable, safe, and reliable operation of stockpiling assets.

Again, it's important to remember that the worst emergency scenarios of a crude oil refinery can occur in stockpiling assets, especially in crude oil storage tanks and LPG tank farms. Figure 7.5 presents the concept of tank level control aiming to avoid overfill occurrences.

7.2 DISCIPLINE

Operational discipline is another key pillar of the safety process in stockpiling operations. Like any operating area of the crude oil refinery, the stockpiling operations have a series of procedures to ensure the safety of the routine operations.

One of the main safety factors in stockpiling operations is the communication quality between the operators involved in the operations. Before starting a transfer operation between tanks, the panel operators should confirm by radio the valves and equipment involved in the operation, aiming to ensure that the planned operation will occur without mistakes.

Communication is only one of a series of procedures and standards which drive the stockpiling operations, and the managers and operation team need to be adequately trained in each procedure and have adequate discipline to comply with the standards without deviations. Figure 7.6 shows what happens when there is a management drift between the planning and the work that is really done.

The model proposed by DEKKER, 2014, says that the deviations are in relation to the procedures and standards.

To avoid drift management over time, it's necessary to train constantly, aiming to keep the operators aware of the risks associated with their day-by-day operations. Unfortunately, the greater the familiarity with the asset, the greater the risks associated with human error in the operation of this asset, once the "familiarity" with a

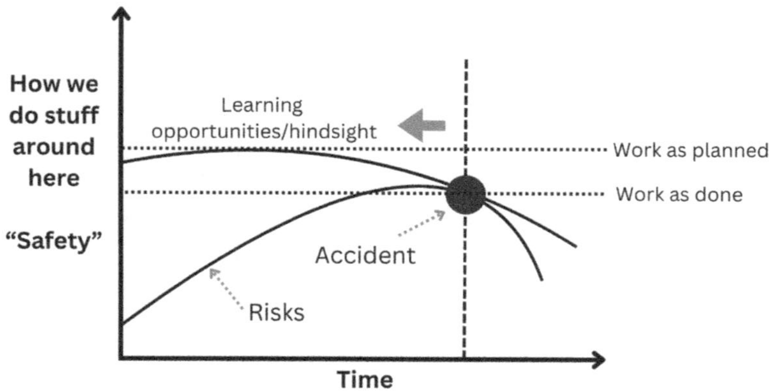

FIGURE 7.6 Effect of management drift between the planning and execution (Based on DEKKER, 2014).

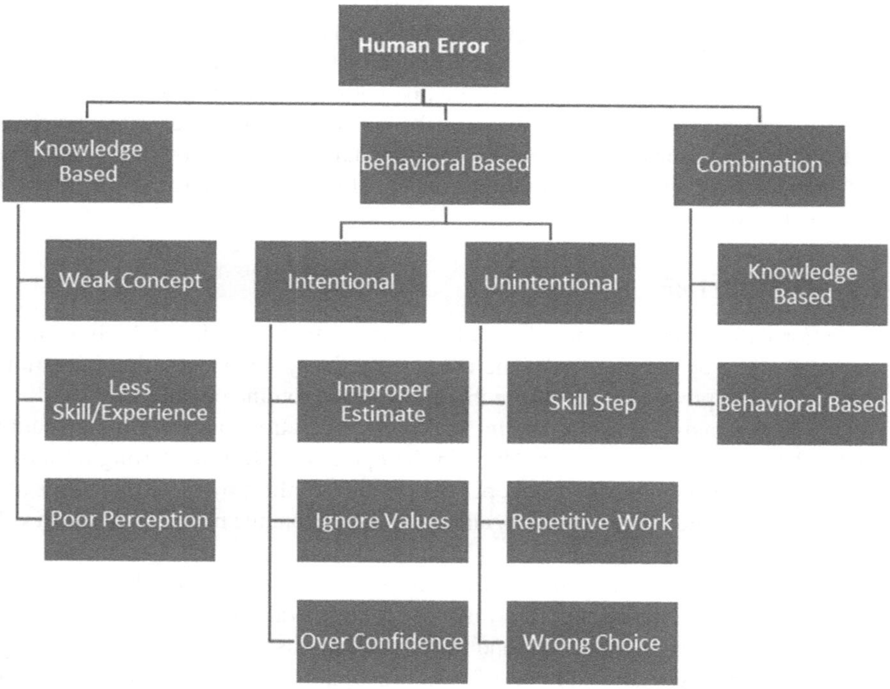

FIGURE 7.7 Classification of human error (Based on REASON, 1990).

refining asset tends to lead to a lack of attention or, on the extreme side, negligence with the risks associated with the operations and the tendency of the team to neglect the procedures and operational discipline. Figure 7.7 presents an overview of the human error classification based on the concept developed by James Reason.

Based on the classification presented in Figure 6.3, it's possible to understand that there is a natural evolutive cycle which demands management actions, aiming to prevent excessive skill from becoming negligence and operational indiscipline, as presented in Figure 7.8.

The management actions described in Figure 7.8 can be related to constant training of the operating team as well as adequate periodic audit planning, aiming to ensure that the operating team is trained and following the procedures without or with minimum deviations. The audit is also an opportunity to identify a lack of knowledge, which can support a training plan.

The audits carried out should cover all the aspects of the operational routine of the stockpiling operators not in order to just find errors and mistakes but to identify procedure deviations, which will reveal the necessity for more and better training programs. Under any circumstances, the audits should be of a punitive nature; the refinery should look to learn from the deviations and take adequate management actions to promote a real just culture in the organization in order to ensure a confident relationship between the management and operation teams.

As described above, one of the key processes related to the operational discipline in stockpiling operations is adequate communication between operators during the transfer of hydrocarbons. An example of how this topic is essential in stockpiling operations is the change of storage tank, which is receiving production from a processing unit. This scenario is especially critical when involving the production of liquefied gases like LPG or propylene.

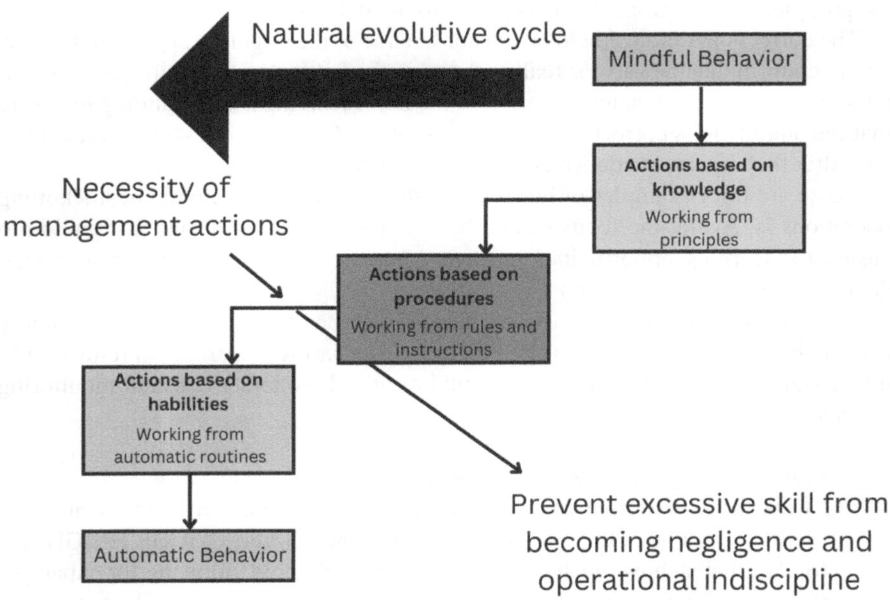

FIGURE 7.8 Natural evolutive cycle between mindful behavior and automatic behavior (Based on REASON, 1997).

In a crude oil refinery, during the change of the LPG production stream from a delayed coking unit from the LPG sphere A to the LPG sphere B, the operator closed the admission valve in sphere A and some minutes before opening the admission valve in sphere B. During this time of interval, the LPG stream alignment was totally blocked, and the process safety valve of the production vessel in the delayed coking unit opened to the flaring system, characterizing a significant process safety incident.

The correct operation in this scenario should be adequate communication between the panel and field operators of the stockpiling area by radio confirming the valve involved in the LPG sphere change, and after the alignment confirmation and valve identification, the field operator should open the admission valve in sphere B, followed by the closing of the admission valve to sphere A. After the confirmation from the field operator related to the adequate sequence of the valve manipulations, the panel operator should confirm that the LPG production is feeding the sphere B through the signal of the gauging system.

In another operational incident, during the change of spheres that received propylene production (from sphere B to sphere A), the panel operator requested the field operator to align the valves of sphere A. The field operator went back to the panel operator, informing them that the production valve of sphere A was opened, but they did not make the operation; they only saw that the valve seemed open, and this was informed to the panel operator. In the sequence, the production valve of sphere B is closed. After around 40 minutes, when the gauge system of sphere A does not indicate flow or level raising, the supervisor of the propylene production unit demands the stockpiling supervisor check the propylene alignment once the pressure in the production vessel is close to the limit. Two minutes later, the pressure safety valve of the propylene production vessel is opened to the flare system.

The correct operation should be the conference of the alignment by the field operator to confirm that the valve is really opened in the field as well, and the panel operator should monitor the gauging system of the new receiving sphere, aiming to ensure that the alignment is correct through the response of the gauging system a few minutes after the alignment confirmation from the field operator.

These are brief examples of how critical the operational discipline in stockpiling operations is. Again, the focus should be ever on identifying deviations, aiming to ensure the learning opportunities in lower-impact events and not under near accidents or in emergency situations.

The transfer operations are among the critical operations in a crude oil refinery and can be responsible for the worst emergency scenarios in a crude oil refinery. For this reason, the transfer operations should be carried out under constant monitoring and adequate procedures as presented below:

- ***Before starting a hydrocarbon transfer:***
  Check the hydrocarbon level of the receiving tank to determine how much product it can take. It is important to determine a SAFE GAUGE HEIGHT (SGH), that is, how much fuel the tank can safely hold, allowing for expansion due to temperature variations. The field operator should walk the pipeline to check for visible leaks or any irregularity in the pipeline as well as to check the valve's position. It is important to make sure that valves are in the proper position so that the product will only go to the targeted tank.

TABLE 7.2
Maximum Ballasting Flow Rate for Storage Tanks (Based on API Good Practices)

Receiving Pipe Diameter (inches)	Maximum Ballasting Flow Rate (m^3/h)
12"	263
14"	357
16"	467
18"	591
20"	730
24"	1050

Determine how much hydrocarbon is above the receiving pipe inside the receiving tank. If there is less than one foot of hydrocarbon above the receiving pipe, the transfer should be carried out at a reduced rate until at least one foot of product is covering the pipe. This will reduce the potential for explosions caused by static electricity when fuel is pumped into the tank at a high rate.

This point is especially relevant for filling tanks backing from maintenance, which needs to be ballasting where the risk of explosions is even more severe. According to the diameter of the receiving pipe, it's possible to determine the maximum ballasting flow rate, as presented in Table 7.2.

If a transfer is made outside of the refinery (for local terminal clients), the stockpiling operator should request an alignment inspection document from the terminal team, which can be sent by e-mail or communicated through the radio, ensuring that the receiving alignment is correct and there is sufficient space in the receiving tank to hold the planned transfer volume. During this time, the operation teams should make contact by radio to check the transfer conditions, but this is especially critical during the start and stop of the transfer, due to the process transients.

- ***During the transfer*:**
  The transfer should start under a reduced flow rate until the operation team ensures that the receiving tank is receiving the transfer without leakages or any deviations. The operation should be accompanied by the operation team all the time, checking the expending and the receiving tank levels periodically.

  The transfer flow rate should be reduced when the safe gauge height is nearing, avoiding overfill risks and allowing sufficient time for the operating team to close valves and perform another procedure.
- ***End of the transfer*:**
  Close the valves of the receiving tank first to avoid hydraulic hammer in the transfer pipe and make the communications to end the transfer operation by checking the valve's position.

  After an adequate time interval (normally 24 hours), the receiving tank can be dipped to check the water level accumulated at the bottom, and then the water can be drained for the appropriate system.

As described above, the operational discipline is fundamental to ensure adequate adherence between how the operations are planned and how they are really carried out by the operating personnel. Again, it's necessary to deal with the deviations in a just way in order to promote a safe environment where the employees trust the management team, allowing that the failures and mistakes will be applied to promote the development of the team and not to look for culprits and generate punishments which lead to a psychologically unsafe work environment. It's impossible to create a real process safety culture in any company without psychological safety, especially in crude oil refineries.

7.3 INTEGRATION

The third pillar of the tripod proposed for the process safety in stockpiling operations in the integration. The term integration is related to a closer relationship between the stockpiling operation team and another operating area of the refinery. It's not uncommon to hear from the stockpiling teams that they were not informed about some operations and that the stockpiling is not adequately involved in the refinery operations planning, mainly in maintenance shutdowns.

This is a serious mistake that can have severe consequences. During planned shutdowns, it's necessary to define the required space for storage intermediates and final products from the process units, which will be kept in operation, and only the stockpiling team has sufficient knowledge about the risks and availability of storage tanks, capable of allowing the safe shutdown and startup of the processing units, including the risks associated with the hydrocarbon stream transfers during these steps once the processing units are in the transient operation regime and the process safety risks are higher.

Another key aspect is the impact of the maintenance shutdown of processing units over the stockpiling area on the discharge to the flare system since the system has a limited capacity, which can bottleneck the processing units shutdown without adequate planning and shutdown sequence. This is especially relevant for refineries located in regions with high population density, since the flare noise, odor, and emissions can lead to reclaims from society in contrast with the current ESG policies from the companies.

It's not uncommon to hear from the processing unit's team that the "culture" of the stockpiling area is much different, leading the stockpiling team to present a kind of "distance" from the other operating areas of the refinery. This must be severely combated by the stockpiling manager. The stockpiling area has specificities, but, at any moment, barriers should be created between the stockpiling area and the processing unit operation. The management team should create incentives and develop a totally integrated environment between the whole refinery, especially the stockpiling, which affects and coordinates any relevant operation in the crude oil refinery.

The stockpiling operation team should be capable of describing which stream is filling and emptying each tank or pipeline, and this is only possible under the perfect integration with the whole refinery.

FIGURE 7.9 Framework of Lencioni model (Adapted from LENCIONI, 2005).

To promote a closer integration between stockpiling and other operating areas of the refinery, the managers can apply some proven integration models like FSPNA and the Lencioni models, aiming to troubleshoot the scenario and build alignment, leading to a better integration. The framework of the Lencioni model is presented in Figure 7.9.

The model proposed by Lencioni establishes the key factor, which led to team failure, as described below:

- Absence of trust: This is especially dangerous for operating teams of crude oil refineries which deal with process safety risks in day-to-day activities, which demand a high confidence environment; the absence of trust can lead the operating teams to hide and not share information, raising the process safety risks.
- Fear of conflict: Teams without confidence cannot find an adequate environment for honest and direct communications explaining their limitations and restrictions, avoiding discussions which normally are necessary for a deep understanding of the scenarios involved in the refinery operations.
- Lack of commitment: The operation teams do not have a complete understanding of their roles, leading to poor engagement with the team and with the whole business, creating a culture of individuality, which can be severely dangerous for the process safety requirements.
- Avoidance of accountability: The team does not develop a team spirit, creating gaps between the positions, creating or worsening risks to the operations and process safety.
- Inattention to results: There is a complete lack of understanding of how the team's role in the crude oil refinery is creating knowledge "islands" between the different areas.

The main role of the Lencioni model is for the managers to understand the weak points of the teams and how to overcome these issues. Considering the context of a crude oil refinery, it's fundamental to promote closer integration and cooperation between operating teams to minimize the impact of non-integrated actions.

The absence of trust is one of the most prejudicial and dangerous characteristics of dealing with personnel teams. This characteristic will lead the operation teams of different areas (mainly the stockpiling team) to adopt a defensive position, aiming to protect their own operations and team, creating then non-integrated operating teams concerned only with their own results and not with the whole results.

The fear of conflict is another dangerous characteristic for stockpiling and refinery operation teams. Conflict is not a bad thing if people are discussing how to develop safer operations in a critical and respectful way. The fear of conflict can lead some operators to not discuss some procedures considered unsafe, leading to opportunity losses to team improvement and learning. The operation team should have the confidence to say when they understand that the procedures and operations need to be reviewed or discuss how an operation can be carried out in a safer or optimized way.

The lack of commitment is caused by the absence of clarity on the role of the team in the business. It's fundamental to create opportunities and forums for the management to clearly describe what the roles of the operation teams are in the success of the refinery. The operators need to clearly understand their impact on the business and how their performance can add or destroy value to the business. This will help to develop commitment in the team, minimizing the risks associated with non-integrated actions.

The avoidance of accountability is a side effect of the absence of trust, leading to each one for himself profile. This will lead to a dangerous scenario where each one will try to protect themselves, which will raise the process safety risks once each operator is concerned with their own safety and not with the whole operation.

Finally, the consequence of all quoted characteristics is the inattention to results once each one is only concerned with protecting themselves against failures and mistakes, prejudicing the results and worsening the process safety risks. To overcome these threats, the complete integration of the crude oil refinery operation teams is essential, as it allows adequate information flow and better decision-making processes capable of anticipating actions and procedures capable of putting under control the inherent risks associated with the refinery operations.

These recommendations are especially applied to stockpiling operating personnel once they are affected by each mistake of the refining unit processing teams, leading to severe risks to the stockpiling assets, like receiving gaseous streams in atmospheric tanks, streams with the temperature above the design temperature of the tank, etc. The stockpiling panel operator and the supervisor should know that each stream is sent for each tank, the operation should be continuously monitored through the alarm system, and a non-identified stream flow to a storage tank should be immediately communicated to the shift coordinator and deeply investigated.

The stopped tank alarm is very helpful in this sense. The panel operator of the stockpiling area should be aware of any changes in the gauge status of the storage tanks, which can indicate unprogrammable transfer between tanks or from processing units to the storage tanks. The philosophy of typical storage tank automation is presented in Figure 7.10 and Table 7.3.

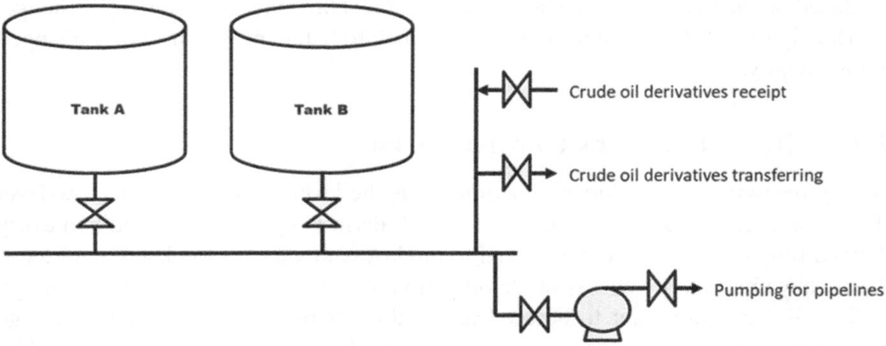

FIGURE 7.10 Typical arrangement for storage tanks receiving and transferring systems.

TABLE 7.3
Storage Tank Operation Status and the Correspondent Intrinsic Risks and Protection Layer

Storage Tank Status	Risk	Cause	Protection Layer
Stopped Empty	Improper Receipt	Fluid passing through the tank valves	High Level Alarm
Receiving Streams	Overfill	Alignments failure	High Level Alarm
	Receive less than expected	Leakage in the alignment	Dynamic low level alarm
Stopped and Filled	Improper Receipt	Fluid passing through the tank valves	High level alarm
	Inventory loss	Fluid passing through the tank valves and/or leakages	Low level alarm
Transferring inventory	Transferring more than expected	Leakage in the alignment	Low level alarm
	Transferring less than expected	Poor pump efficiency, improper alignment blockage	Dynamic high level alarm

7.3.1 High- and Low-level Alarm

At the end of product movement (receipt or transferring), the level is expected to remain unchanged until the next operation.

The reference level is then changed to the first level value after termination.

Two level difference alarms are created: high and low, taking the previous value as a reference.

7.3.2 High- and Low-level Alarm

When starting a receiving or shipping operation, the operator defines the volume to be pumped.

Based on the diameter of the tank, it is calculated how much the level will change.

The high-level (for receiving) or low-level (for transferring) alarm is then changed automatically.

7.3.3 Dynamic High- or Low-level Alarm

During receiving or transferring operations, as the level rises or falls, the low-level (for receiving) or high-level (for transferring) alarm is dynamically modified every defined time interval in seconds, equal to the last read value of the level, in order to that, if any abnormality occurs in the operation, the alarm is triggered immediately.

This is very important to detect leaks and dangerous transfers for the storage tanks, which can lead to severe process safety incidents or accidents, as described by the following examples.

During an operational instability that occurred in the crude oil distillation unit of a crude oil refinery, the debutanizer column, which is responsible for separating the LPG from the straight-run naphtha, as presented in Figure 7.11, lost performance, and the high vapor pressure naphtha (naphtha with high LPG concentration) was kept aligned to the naphtha hydrotreating unit storage tank.

The stockpiling operation team personnel was not communicated, and the mixture of naphtha and LPG was kept aligned with Tank A (naphtha hydrotreating feed tank). After approximately 30 minutes, the feed pump of the naphtha hydrotreating unit suffers cavitation, leading to the emergency shutdown of the processing unit.

At the first moment, this incident can be faced as a simple emergency shutdown of a processing unit, but it's necessary to understand that nothing is simple in a crude

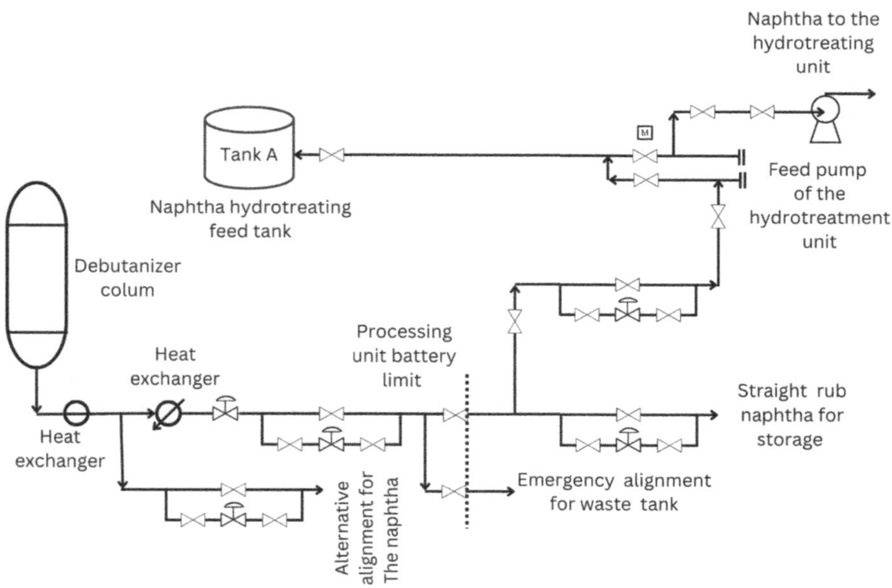

FIGURE 7.11 Process arrangement for straight run naphtha system of a crude oil refinery.

Scenario	Risk	Consequences	Mitigation Actions
• Operational Instability in the Debutanizer Column.	• Sending High Vapor Pressure Naphtha to the storage tank.	• Tank overpressure; • Cavitation of the feed pumps of the naphtha hydrotreating unit.	• Comunicate immediately the stockpiling team; • Changing the naphtha alignment to residue tank or to the flaring system.

FIGURE 7.12 Example of SIMOPs application.

oil refinery. Fortunately, the pump stopped under cavitation, and the only consequence was the shutdown of the processing unit. The pump seal could have failed and led to the contention loss of LPG and naphtha with great potential of fire. Furthermore, the startup of a hydrotreating unit is not a simple task and exposes the operators to risks. It's important to understand that the most severe risk scenarios in a crude oil refinery are during the transient moments like the shutting down and startup of processing units and other operational systems.

A SIMOPs analysis could help the refiner to take actions to avoid this process safety risk scenario through an adequate communication process between the two operation teams and propose actions capable of interlocking the process plant, avoiding the alignment of the high vapor pressure naphtha to the storage tank (sending it to a residue tank or flaring system during the operational instability). Figure 7.12 shows an example of the result of the SIMOPs application in this case.

Another example of integration failure between processing units and stockpiling operation teams occurred in a crude oil refinery during the startup of the fluid catalytic cracking unit (FCCU).

During the operational routine, the stockpiling field operator identified the crude oil projected over the floating roof; the tank was improperly receiving the high vapor pressure naphtha from the FCC unit, which is under the starting up procedure. The crude oil storage tank doesn't have sufficient hydrocarbon levels to absorb the LPG, leading to the crude oil projection over the roof through the roof seal.

During the root cause analysis, it was discovered that the FCC operation team was facing difficulties heating the bottom of the debutanizer column of the processing unit, leading to the production of high vapor pressure cracked naphtha (cracked naphtha with high LPG concentration); again, this was not informed to the stockpiling operation team, and the stream was kept to a crude oil storage tank with a low hydrocarbon level. The contention loss of the crude oil could lead to fire and to one of the most severe emergency scenarios of a crude oil refinery, which is fire in crude oil storage tanks; another consequence could be the loss of the crude oil storage tank due to the roof sink once the buoyancy density limit should be reached.

These examples reveal the severity of the integration between the operating teams in a crude oil refinery and the process safety requirements of the stockpiling area.

Any changes in the process should be communicated to the stockpiling team, aiming to allow operating procedures to mitigate the risks and minimize the impact of instability, leading to safer operations.

The adequate integration between the operation teams of a crude oil refinery depends on the construction of a safe and just culture environment where the team can develop confidence in the management team and between each team member, as presented above. This can be reached through an adequate integration management policy, which must be encouraged and led by the stockpiling manager once they have (or should have) an adequate view of all processing units and the complete understanding that any change in the processing units will impact the stockpiling assets under your management. As previously mentioned, most of the stockpiling and hydrocarbon transfer operations involve simultaneous operation between at least two different operation areas and should be conducted as SIMOPS studies for each operation characterized as critical by the stockpiling operation team, mainly those associated with the startup and shutdown of processing units, which will lead to transient flow to storage tanks.

BIBLIOGRAPHY

American Petroleum Institute, API Standard 2350, *Overfill Protection for Storage Tanks in Petroleum Facilities*, API Standard 2350, Draft 5th Ed., April. 2021.

American Petroleum Institute, API Standard 537, *Flare Details for General Refinery and Petrochemical Service*, API Standard 537, Draft 3rd Ed., Sept. 2001.

American Petroleum Institute, *Guide for Pressure-relieving and Depressing Systems*, API Recommended Practice 521, 6th Ed., January 2014.

Baybutt, P. *Simultaneous Operation (SIMOP) Review: An Important Hazard Analysis Tool.* Process Safety Progress, Volume 36 - Issue 1, Wiley, 2017.

Center for Chemical Process Safety (CCPS), *Guidelines for Risk Based Process Safety*. 1st ed. Wiley, 2007.

Chang, J. I.; Lin, C.C. *A Study of Storage Tanks Accidents*, Journal of Loss Prevention in the Process Industries, Elsevier, 2006.

Dekker, S. *Just Culture – Restoring Trust and Accountability in your Organization*. 3rd ed. CRC Press, 2017.

Dekker, S. *The Field Guide to Understanding "Human Error"*. 3rd ed. CRC Press, 2014.

John Zink Co., *The John Zink Hamworthy Combustion Handbook*, Vol.3, Applications, Chapter 11: Flares, CRC Press LLC, 2001.

Lencioni, P. *The Five Disfunctions of a Team*. 1st ed. Wiley, 2005.

Reason, J. *Human Error*. 1st ed. Cambridge University Press, 1990.

Reason, J. *Managing the Risks of Organizational Accidents*. 1st ed. Routledge Press, 1997.

8 Calling the Attention to Transfer and Stockpiling Area

The Relation with Maintenance

Despite the relevance of stockpiling assets to the process safety, reliability, and profitability of the crude oil refineries, it's not uncommon to identify refineries where the maintenance and asset management policy of stockpiling assets are underappreciated in relation to refining units.

The robustness of stockpiling assets leads to the false impression that the maintenance of these assets can be put under secondary control in comparison with the refining units once the impact on production tends to be immediate. Another factor which contributes to the low priority of the maintenance of stockpiling assets is that these operating systems do not have planned shutdowns for maintenance like the processing units, as well as the high capital demanded for adequate management of these assets, which can lead the refiners to use the maximum lifecycle of these assets until the maintenance for a complete maintenance, especially for the storage tanks.

In some cases, the immediacy of routine maintenance and certain management myopia that do not allow for an adequate sense of urgency in relation to the transfer and stockpiling of assets contribute to a low prioritization of resources for these assets.

It's fundamental to understand that the key performance indicators (KPIs) operation and maintenance areas in a crude oil refinery tend to be antagonistic and the stockpiling manager should be able to overcome this barrier to call attention and maintenance resources to establish an adequate asset management policy, aiming to ensure process safety requirements, reliability, and minimum impact over the production planning.

Due to the immediate impact on the production planning of the refinery, the routine maintenance resources tend to be displaced to the refining units to the detriment of stockpiling assets, which present less potential to cause an immediate impact on the production planning. Unfortunately, the stockpiling of assets, as well as any operating system, has an operating lifecycle that will be reached; for the processing units, this normally coincides with the maintenance shutdown that does not exist for the stockpiling assets (at least in the same way).

DOI: 10.1201/9781003472216-8

To overcome these difficulties, it's necessary to establish adequate asset management policies, especially for the stockpiling assets, aiming to ensure that the stockpiling assets will not compete for maintenance resources with the refining units. This assumption is even more relevant for storage tank maintenance, which requires great capital investment and a high amount of resources. In this case, a dedicated maintenance workforce should be applied.

The stockpiling manager should be able to attract resources and attention to the stockpiling area, and for this, they should have a deep knowledge of the refinery processes. If possible, the stockpiling manager can know in detail the daily production planning, aiming to identify the bottlenecks and anticipate actions and maintenance capable of avoiding or minimizing the impact of equipment failure or restriction over the production planning.

Through a deep knowledge of the business, the stockpiling manager can expose the potential impact of the stockpiling assets on the refinery operation and awaken the adequate sense of urgency of the maintenance team to the stockpiling assets.

Before starting an adequate maintenance plan for stockpiling assets, it's necessary to understand the role of the maintenance team in a crude oil refinery. The maintenance team is responsible for repairing failed equipment and systems as well as adequate management actions capable of ensuring preventive actions to minimize the occurrence of failures, aiming to avoid process safety risks and impacts over the production planning.

It's a common issue in crude oil refineries: the overload of the maintenance team with upgrades and changes in the operating systems, which displace the maintenance resources from routine maintenance to operating improvement actions, contributing to the reliability and integrity degradation of the process plant.

It's fundamental to any refinery to rely on engineering or enterprise teams which will be responsible for change implementations through an adequate change management program. This is a key topic for process safety management in crude oil refineries; once an inadequate change management can lead to severe process safety accidents, for this reason, the change management should be conducted by a multitasking team allowing a deeper understanding of the change and adequate actions to mitigate eventual risks. This is one of the reasons that the changes and upgrades of crude oil refining systems shouldn't be conducted only by maintenance teams which normally do not have the total knowledge depth to carry out the change management.

Considering the role of the maintenance team in the crude oil refinery, it's necessary to define an adequate asset management policy. As previously mentioned, the stockpiling assets demand special maintenance plans and asset management policies in order to avoid the trend of higher urgency sense by the managers in relation to the processing units like FCC and delayed coking and reduce the available workforce to keep the reliability and integrity of the stockpiling assets.

Normally, the traditional maintenance is classified in four categories:

- Repair: In this case, the equipment or system is operated until the failure, and there is not an adequate asset management policy. This maintenance policy is called corrective maintenance.

- Condition-based: The health of the operating asset is not monitored, and maintenance is scheduled when the equipment or systems indicate poor performance or failure signs.
- Scheduled: The maintenance is scheduled at periodic times based on the history of the equipment.
- Reliability centered: These are adopted management actions capable of anticipating equipment and systems failures or poor performance in order to ensure a lower probability of impact over the production planning or, even more important, over the human and process safety. The reliability-centered maintenance can be implemented under different policies, as presented in Figure 8.1.

As presented in Figure 8.1, preventive maintenance can be an attractive way to control the reliability and availability of production assets. This policy is based on fixing equipment and systems before a failure, but this can lead to capital loss with shorter operating lifecycles or the risks associated with failures and production losses due to failures caused by too-late maintenance. Considering the characteristics of the stockpiling assets, preventive maintenance can represent even more inadequate once failures can lead to severe process safety accidents and production losses, and the earlier maintenance can represent great capital costs; this is especially true for the storage tanks.

The modern approach of maintenance policies is based on data-driven in order to identify failure trends, aiming to allow adequate time for taking actions to avoid or minimize the impacts of the failure.

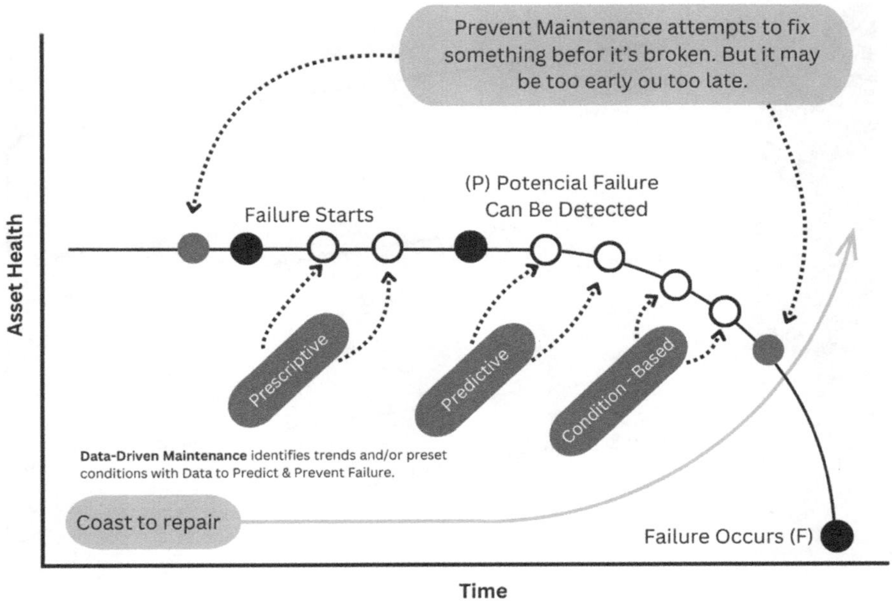

FIGURE 8.1 Asset management policies.

Predictive maintenance is applied through periodic monitoring of the equipment using data like vibration or corrosion rates to predict the failure and allow the maintenance before the failure occurrence.

Prescriptive maintenance was developed based on the concepts of machine learning, IoT (Internet of Things), and digital transformation, aiming to allow constant monitoring of asset health. Prescriptive maintenance allows the optimization of asset performance, load balancing, and production planning actions based on asset health.

It's important to highlight that the major challenge in digital transformation is precisely to convert the data into useful information which supports an agile and efficient decision-making system. It's a misconception that digital transformation is purely a technological event. Technology allows easy and quick access to large quantities of data; however, real digital transformation consists of a change of mind map in the sense of integrating, making critical analysis, and finally converting this data into actions that lead to improvement of the decision-making process.

Presenting the data in a simple way, with the agility of acquisition and space for the insertion of critical analysis, which should be treated as another integrated data, will allow better performance of the prescriptive maintenance policy. The data, presented in an integrated and easy reading method, together with the analysis carried out by the technical staff, are transformed into information that leads to more clarity in the decision-making process by the management staff, always respecting the concept of maintaining the integrity of current operations, minimizing losses, and developing the bases to the refinery strategic plan. Figure 8.2 presents an example of the

FIGURE 8.2 Decision-making process flow based on data.

decision-making process flow which can be adopted through a management system based on data.

One of the key questions to achieve the desired results is an adequate asset management system. Due to the limitation of maintenance resources, one of the main deliverables of the project is the prioritization of maintenance orders that need to be accomplished, keeping the focus on adding security and profitability to process units. The combination of real time monitoring of process data and the maintenance prioritization system allows faster identification of inefficiency points and higher agility to the solution, adding value to the business.

The application of an asset management policy based on prescriptive maintenance for stockpiling assets allows the refinery to minimize the production losses due to equipment failure, especially for dynamic equipment like pumps and storage tank mixers. Prescriptive maintenance can also be used to monitor and manage the operating lifecycle of storage tanks using IoT devices to measure and monitor the corrosion rates, allowing one to determine the best moment to put the tank out of service and start the general maintenance, even before the legal time intervals.

As previously mentioned, due to their characteristics (expensive maintenance, longer time frame for impact over the production planning, and management myopia), there is a trend that the stockpiling area loses priority in relation to processing units regarding the maintenance resources. To overcome this scenario, the stockpiling manager needs to have adequate resilience, technical and management knowledge, and capacity to deeply understand the complex scenario of a refinery, aiming to expose how the stockpiling assets and their failure or unavailability will impact the refining business.

With a deep understanding of the refining business, the stockpiling manager will be able to expose the relevance of the stockpiling area to the refinery value chain and call attention and maintenance priority to their area. An attractive way to achieve this goal is to use negotiation techniques.

As previously mentioned, stockpiling assets tends to be underestimated in relation to the refining processing units' assets during the maintenance prioritization. The stockpiling management team should be able to find adequate ways to describe the relevance and potential impacts of stockpiling asset management failures over the production planning and, even more relevant, over the process safety requirements of the whole refinery.

A good way to ensure the adequate prioritization of stockpiling assets is to apply negotiation techniques. To ensure success in this strategy, the stockpiling manager should have adequate knowledge of the whole crude oil refinery as well as be integrated with the other production areas. Furthermore, it's fundamental to have a deep understanding of the production planning, aiming to identify bottlenecks and how the failure of the stockpiling assets will impact the refinery operation and profitability.

It's fundamental to understand that the relationship with maintenance and operation teams tends naturally to be conflictive, and this is not always a bad thing. Sometimes, conflict is necessary to find a real and effective solution. In mature organizations, conflict can lead to an increase in a team's commitment and loyalty as well as innovative breakthroughs and new approaches to solve problems. On the other hand, immature organizations conflict can cause frustration, fear of failure, and lower productivity with wasteful conflicts.

The main cause of the conflicts between the maintenance and operation teams is the existence of opposite interests. The maintenance team tends to keep their focus on the maintenance costs, trying to minimize the services, while the operation team will demand the maximum availability of the refining hardware with the maximum reliability, which normally means raising the maintenance costs.

Considering the trend of the stockpiling assets receiving less priority than the refining processing units, a potential conflict in the crude oil refinery is between the stockpiling manager and the other operation managers due to the competition over the scarce maintenance resources.

According to the literature, there are five main negation modes to respond to a conflict:

- Competitive: In this case, satisfying the needs of the other party is not important to the negotiator and establishes a win-lose relationship. This negotiation method should be avoided, as it will cause more conflicts and revenge feeling between the involved parties.
- Collaborative: Satisfying the interest of both parties is important during the negotiation process and establishes a win–win relationship. This is the best negotiation way to create an environment of mutual confidence.
- Avoiding: The negotiator is indifferent about satisfying either their own needs or the needs of the other party;
- Accommodating: When the negation target is attractive just to one part, and the conflict can be solved with a simple yield.
- Compromising: Satisfying the interest of both parties is moderately important.

In most companies, and this is not different in crude oil refineries, the most common negotiation profiles are the collaborative and the competitive profiles.

As previously described, the competitive or win–lose profile should be avoided, as it will lead to resentment and normally create even more conflicts. This does not mean that competitive negotiation should never be applied. In some cases, this is the only way to follow, but it's necessary to consider the consequences and take adequate actions to minimize or neutralize these consequences.

As previously mentioned, the relationship between the operation and maintenance personnel tends to be conflictual due to the antagonism of the KPIs which are applied to measure the performance of each team, and this tends to be even more relevant considering the stockpiling operation team once there is a trend of the stockpiling assets losing maintenance priority in relation to the processing units assets. Unfortunately, the relationship between the operation and maintenance team does not receive great attention from the companies with great impact on both parts and the whole company.

Through a good relationship between the operation and maintenance team, it's possible to reach a confident environment, avoiding the conventional tendency that teams must protect themselves and avoiding an adequate information flow that leads to the identification of root causes of equipment failures, for example. This helps to reach a better strategy to prevent breakdowns and prioritize equipment maintenance based on the value generation to the refinery.

To understand and manage adequately the relationship between the operation and maintenance team in a crude oil refinery, it's fundamental to understand or remember the role of the crude oil refining business. A crude oil refinery should generate value for society through adequate management of their processes, ensuring process safety, environmental protection, and profitability.

The main goal of a crude oil refinery is to maximize the added value to the processed crude oil by applying the safest routes possible. It's always important to make clear that the crude oil refinery does not produce maintenance, and the role of maintenance is to support the operations of the refinery, being a service provider to the operation.

A good way to improve the quality of the relationship between maintenance and operation is to create an adequate information flow, taking some actions:

- Make clear the needed routine inspections and the basic care of the equipment.
- Define together the criteria of maintenance priorities, aiming to manage the expectations of both parts.
- Make clear the production plans, aiming to allow adequate time to identify eventual threats and equipment restrictions. This will allow adequate time for the maintenance team to act to avoid production losses, reducing the exposure of the maintenance. This will help to establish a confident relationship between maintenance and operation and create space for a win–win negotiation.

The actions quoted above can be summarized as "make the operation and maintenance teams to talk."

As described above, the relationship between the stockpiling operation and maintenance teams tends to be hard, but this relationship is essential to avoid process safety risks and production losses in the crude oil refinery. The adequate synergy between these teams is fundamental to the results of the refinery, and the stockpiling manager should be engaged to find ways to adequately call attention to and resources from the maintenance team for the stockpiling area. This can be reached through an adequate negotiation approach with the maintenance managers as well as the definition of assertive KPIs, aiming to ensure that the key reliability and asset integrity topics of the stockpiling area will be adequately managed by the maintenance and inspection teams.

The stockpiling manager should conduct frequent meetings with the maintenance teams, aiming to present the main concerns and verify the compliance of the maintenance plans as well as define new maintenance strategies to minimize the asset integrity degradation.

In summary, the relationship between operation and maintenance in a crude oil refinery tends to be hard, especially regarding the stockpiling area, which tends to receive less priority from the management system than the refining units.

The stockpiling manager needs to use their operating and process knowledge to find adequate ways to call attention to the stockpiling assets, demonstrating that despite the lesser impact of the stockpiling assets in the short term over the refinery production plan, the consequences of an eventual failure will present greater impacts than the production units. As previously mentioned, it's important to take into account that a conflict will not be a good strategy for any stakeholder, but, if necessary, the stockpiling manager needs to sustain your position.

Changing the culture is always a hard task in any company, and this is no different in a secular industry like the crude oil refining sector. It's impossible to change the culture by making management in the same way, and, fatally, the culture change will demand a certain degree of conflict once the people will tend to sustain their positions and comfort zone. Again, the stockpiling manager should be able to overcome this challenge through an adequate priority sense, management, and technical knowledge, as well as character strength to make the things which really need to be done.

In summary, the transfer and stockpiling manager must have the ability to use the management system to force maintenance to prioritize the management of assets in his area. To do this, he must have in-depth knowledge of the refinery and, most importantly, be continually updated on operational issues and the main threats to transfer and storage in relation to the refinery's production planning. Through this, he will be able to anticipate actions aimed at eliminating these bottlenecks. If necessary, the impact on production must be allowed to draw attention to the need to prioritize resources for asset management in the transfer and storage area.

BIBLIOGRAPHY

Cann, G.; Goydan, R. *Bits, Bytes, and Barrels – The Digital Transformation of Oil and Gas*. 1st ed. MadCann Press, 2019.

Gelfand, M. J.; Brett, J. M. The *Handbook of Negotiation and Culture*. 1st ed. Stanford Business Books, 2004.

Senge, P. M. *The Fifth Discipline: The art and practice of the learning organization*. 2nd ed. Random House Books, 2006.

Silva, M.W.; Laidens, C.P. *Brazilian Refinery Digital Transformation Focuses on People over Process*. Oil & Gas Journal, 2020a.

Silva, M.W.; Laidens, C.P. *Digital Transformation Improves Brazilian Refinery Operations*. Oil & Gas Journal, 2020b.

Sutton, I. *Plant Design and Operations*. 2nd ed. Gulf Professional Publishing, 2017.

Silva, M. W. Personal Notes.

9 Asset Integrity Management in Transfer and Stockpiling Sector

One of the major challenges to managing stockpiling areas is to deal with very big assets such as kilometers of pipelines, hundreds of storage tanks, and pumps which in contrast with the refining processing assets do not have planned shutdown windows. Under this scenario, the main way to ensure the process safety requirements as well as adequate performance with minimum production losses is to ensure an adequate asset integrity plan aiming to maximize the operating lifecycle of the assets balancing the maintenance costs and process safety risks, mainly related to hydrocarbon contention losses.

To achieve this goal, it is fundamental that the stockpiling manager carry out a closer integration with the equipment inspection area. Despite the existence of the equipment inspection team which is technically responsible for the integrity of the whole refinery asset integrity, it's important to understand that only the operation personnel have adequate knowledge of the process characteristics. These characteristics can demand more attention in the care of the processing assets such as salt deposition, dead legs, process fluids with higher corrosion rates, etc. Furthermore, the operating personnel have adequate knowledge of the consequences of contention loss and process safety accidents in the operating assets, mainly in hydrocarbon storage tanks.

In other words, as previously described in the section about the relationship between operation and maintenance teams, it's fundamental to reach a confident and mutual cooperation relationship between the stockpiling operation and the equipment inspection teams. Of course, this is true for all operation areas of the refinery, but due to the characteristics of the stockpiling assets, this is even more relevant to the stockpiling area.

The main threat to the asset integrity of stockpiling assets (mainly pipelines and storage tanks) is the corrosion process. Unfortunately, corrosion is a common and deep concern to the crude oil refining industry which needs to be adequately managed aiming to ensure sustainable management based on the three Ps of the sustainability tripod: People, Planet, and Profit, which need to be prioritized in this order aiming to prioritize the life and environment in detriment of the business if necessary. As previously mentioned, asset management is a pillar of the process safety policy in stockpiling operations and the corrosion threat needs to be adequately managed through the assertive asset integrity plan.

DOI: 10.1201/9781003472216-9

9.1 THE CORROSION THREAT – A BRIEF REVIEW

The corrosion process can be defined as the degradation of metal properties such as mechanical resistance, being the metal converted to a non-metallic state by a chemical or electrochemical action. The most common corrosion process is aqueous corrosion which demands the presence of water, and the corrosion process is based on electrochemical reactions.

The corrosion reactions involve basically anodes (the corroded metal), cathodes, known as electrodes, and electrolytes which are the solution through which the ions are transported from the anode to the cathode. It is important to note that the electrodes can be two different metals or distinct areas of the same metal. During the corrosion process, two reactions occur: anodic reactions in the anode which is characterized by electron loss, and the cathodic reaction which involves the reduction of the electrolyte ions which consume the electrons released in the anode. Figure 9.1 presents the basic concept of electrochemical corrosion.

Considering iron corrosion as an example, when the iron is corroded is produced positively charged ion in the anodic reaction as presented below:

$$Fe \rightarrow Fe^{2+} + 2e^-$$

The electrons produced from the reaction move through the metal to another location where they are in turn consumed in a reaction that produces hydroxyl ions. The reaction depends on the nature of the electrolyte, but typically is one of the following:

$$O_2 + 2\,H_2O + 4e^- \rightarrow 4OH^-$$

$$2\,H_2O + 2e^- \rightarrow H_2 + 2\,OH^-$$

The reactions represented above are referred to as cathodic reactions. The movement of the ions through the electrolyte completes the electrical circuit. The iron ions typically react with the water and/or oxygen to form a corrosion deposit of rust or some

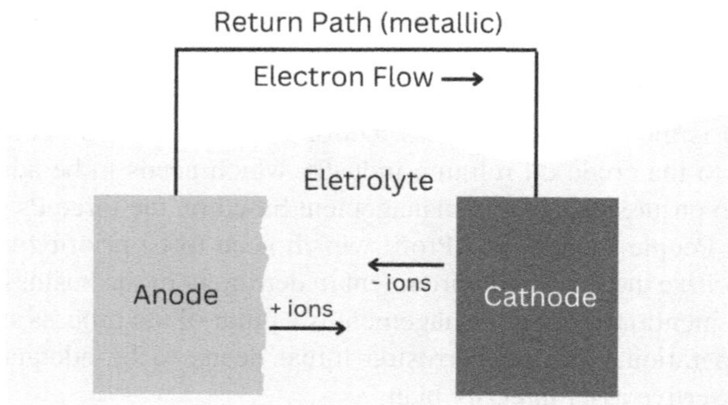

FIGURE 9.1 Typical electrochemical cell.

other iron oxide, but, in some cases, they may react with carbon dioxide or hydrogen sulfide to form iron carbonate or iron sulfide.

9.1.1 Corrosion Classification and Forms

The corrosion process can be classified based on several factors such as electrolyte, corrosion product, and mechanical factors. Figure 9.2 summarizes the classification of the corrosion processes.

Corrosion can occur in different ways, and its classification can be made based on the appearance of the corroded metal.

Corrosive processes of an electrochemical nature present identical mechanisms because they will always consist of anodic and cathodic areas, between an electron current and an ion current circulation. However, the loss of mass and mode of attack on the material occurs in different ways.

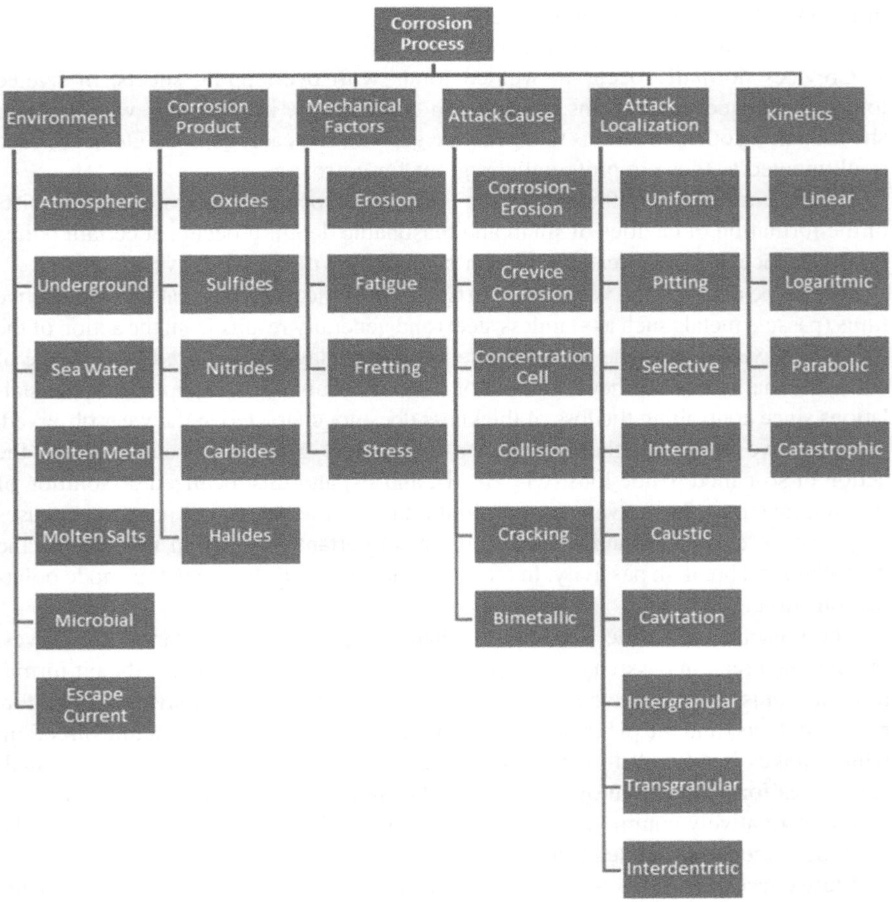

FIGURE 9.2 Classification of the corrosion process.

Uniform corrosion – It consists of the attack of the entire metal surface in contact with the corrosive medium with the consequent decrease in thickness.

This type of corrosion generally occurs due to micro piles of local action and is probably the most common type of corrosion, especially in corrosive processes of structures exposed to the atmosphere and other media that result in a uniform action on the metal surface.

Uniform corrosion is a form of wear that is easier to monitor, especially when it comes to internal corrosion in equipment or installations, considering that the loss of thickness is approximately the same across the entire metal surface.

However, it is an important type of corrosion from a wear point of view and can lead the equipment or installation to significant failures, limiting its useful life. Other types of corrosive attack where there is a preferential location for corrosion to occur, resulting in a localized loss of thickness, are called localized corrosion.

Crevice corrosion – Crevices are subject to the formation of piles of differential aeration and differential ionic concentration. When the medium is liquid, piles of differential ionic concentration preferably occur and when the medium is gaseous, piles of differential aeration tend to occur.

Crevices normally occur in welded joints with overlapping sheets, in riveted joints, in flanged connections, in threaded connections, in coverings with screwed sheets, among other situations that generate gaps. In any case, gaps should be avoided or eliminated as they are preferential regions for corrosion.

Pitting corrosion – Pitting corrosion is a form of localized corrosion that consists of the formation of cavities of small and reasonable depth. It occurs at certain points on the surface while the rest can remain practically without corrosive attack.

It is a type of corrosion very characteristic of metallic materials that form protective films (passive metals such as stainless steel) and generally results from the action of the active–passive island at the points where the passive layer is ruptured. It is a type of corrosion that is more difficult to monitor when it occurs inside equipment and installations since controlling the loss of thickness does not characterize the wear observed.

In passive materials, the breakdown of passivity generally occurs through the action of so-called halide ions (Cl-, Br-, I-, and F-) and this localized dissolution of the film generates an active area that, in the face of the remaining passivate, causes very intense corrosion, and is localized. An important quantity in this case is the potential for a break in passivity. In fact, what happens is a change in the anode polarization curve.

The presence of halide ions causes changes in the anodic polarization curves, making the break in passivity more likely. Another important aspect is the pit formation mechanism, as failure begins at points of weakness in the passivating film (formation defects) and the pH inside the pit changes substantially in the acidic direction, which makes it difficult to restore the initial passivation. It follows that the small active area formed in front of a large cathodic area causes intense and localized corrosion. A relatively common case of passivating breaking is the stainless steel in the chloride presence which leads to pitting corrosion.

Plate corrosion – This is a type of localized corrosion that is characterized by the formation of the plate by the corrosion products which progressively detached, exposing the metal to a new attack.

Galvanic corrosion – Galvanic corrosion occurs when a metal is put in contact with the noblest material in the presence of an electrolyte, leading this metal to act as an anode. The galvanic series of metals and alloys is used to predict if this type of corrosion can occur when different metals are put in contact.

Intergranular corrosion (IGC) – The corrosion process occurs in the boundary of material crystallites.

Transgranular corrosion – Transgranular corrosion occurs when a crack does not follow material grain boundaries but travels right across the grain.

Concentration cells corrosion – Concentration cells are a form of galvanic corrosion. Just as two dissimilar metals joined cause corrosion, so do dissimilar conditions within the electrolyte. Corrosion occurs when two or more areas of the same metal surface are in contact with electrolytic solutions of different concentrations. The same metal has different electrical properties in the presence of different concentrations of the same electrolyte.

Differential aeration (oxygen concentration cell) and ion concentration (salt concentration cells) create dissimilar polarities (anodic and cathodic areas). Differences in dissolved oxygen concentration led to localized corrosion of metal in hidden areas such as under deposits or in crevices.

Corrosion associated with fluid flow – In fluid flow, corrosive processes can be accelerated due to the association of the mechanical effect with the corrosive action. The main types of corrosion associated with a flow are corrosion–erosion, cavitation corrosion, and turbulence corrosion. Erosion of metallic material is the mechanical wear caused by the surface abrasion of a solid, liquid, or gaseous substance. This process is common when the fluid presents solids that can clash with the pipe removing the protective film that is produced during the corrosive process, exposing the metal surface, and accelerating the corrosive process. Figure 9.3 presents an example of the occurrence of corrosion–erosion in tubulation.

The corrosion–erosion process is a special concern for the integrity of stockpiling assets. In the off-site design step of a crude oil refinery, special attention should be applied to define the maximum velocity in the pipelines aiming to avoid the corrosion-erosion phenomena, mainly for liquids that can contain solids particles which will raise the severity of the issues caused by corrosion-erosion process.

Cavitation corrosion occurs in low-pressure regions, which leads to the collapse of gas bubbles, during this process, the protective film over the material is removed leading to the acceleration of the corrosive process. This process is commonly

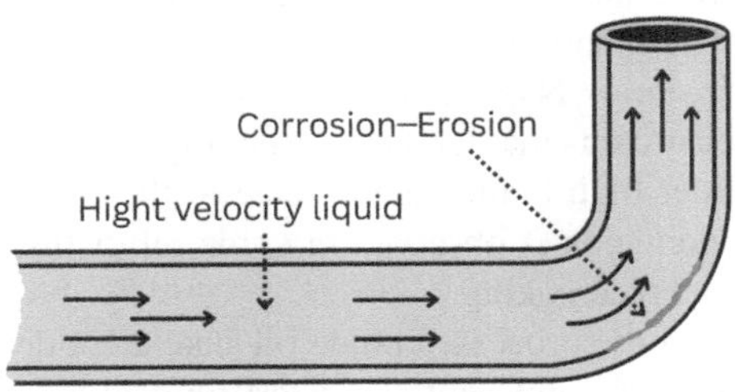

FIGURE 9.3 Corrosion–erosion phenomena in a tubulation.

associated with the pump operation which is a daily operation in stockpiling areas of crude oil refineries, as presented in Chapter 2. The required net positive suction head (NPSH) should be attended to avoid cavitation phenomena.

Meeting the required NPSH ensures that the fluid has sufficient energy in the pump suction to be kept in the liquid phase, avoiding the vapor formation, and consequently eliminating the cavitation risk.

Selective corrosion – Corrosive processes called selective corrosion are those in which a galvanic couple is formed due to the large difference in oxidation potential between two elements of a metal alloy. The most common types of selective corrosion are graphitic corrosion and dezincification.

Graphitic corrosion – Corrosive process that occurs in gray cast iron and nodular cast iron. Cast iron is normally used for water, sewage, and drainage pipes, among others. Since graphite is a much more cathodic material than iron, the graphite veins, or nodules in cast iron act as a cathodic area while iron acts as an anodic area, transforming into a corrosion product. To minimize graphitic corrosion problems, it is common practice to coat the pipes.

Corrosion by dezincification – Dezincification is the corrosive process observed in zinc alloys, especially brass, used in heat exchangers (coolers, condensers, etc.) and piping for salt water, for instance. The corrosion process results in the destruction of zinc (the most anodic material), leaving behind copper and corrosion products. A greater tendency to dezincification is observed in brass with a high zinc content, dezincification can be avoided by heat treating the alloy to solubilize it and by using alloys that contain inhibitory elements such as arsenic (As) and lead (Sb).

Corrosion under insulation (CUI) – This is a relatively common corrosion mechanism that is found in stockpiling assets. As described by the name, this corrosion process is observed in tubulations and equipment like tanks which conduct and store hot products like asphalt and vacuum residue where the service temperature demands the use of insulation both to keep the internal temperature and protect the operating and maintenance personnel against injuries. The corrosion under insulation occurs when there is moisture concentration over the insulation on the external surface; the process is most severe in equipment that operates under cyclic conditions once there is temperature flotation leading to the accumulation of moisture in the insulation which is kept in contact with the equipment surface, raising the corrosion rates. Once the equipment surfaces are not generally accessible for visual inspection, the onset of corrosion cannot be easily identified, and in extreme cases, severe corrosion with consequential loss of system integrity can occur leading to contention loss and process safety accidents. The API 583 presents the main strategies to deal with corrosion under insulation and fireproofing.

Naphthenic corrosion – Naphthenic corrosion occurs in processing units that process bottom barrel streams that tend to present high concentrations of naphthenic acids and operate under high temperatures. Due to these characteristics, the naphthenic corrosion phenomenon is observed in crude oil distillation units and residue upgrading units like delayed coking.

The characteristics of the processed crude oil slate are a determinant factor in the naphthenic corrosion. A very relevant characteristic of oils for refining hardware is naphthenic acidity. Naphthenic acidity is determined based on the amount of KOH

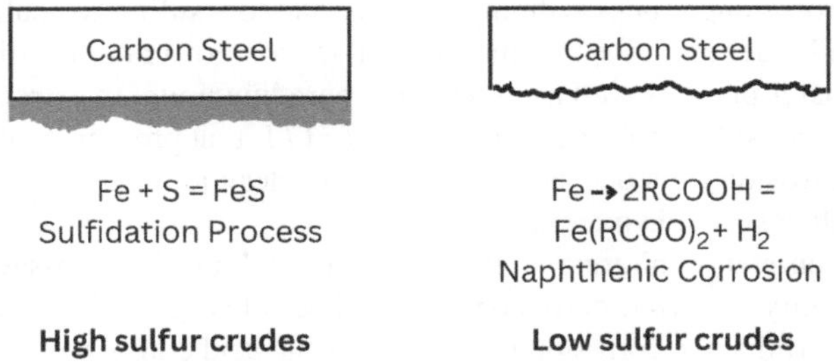

FIGURE 9.4 Naphthenic corrosion process.

required to neutralize 1.0 gram of crude oil, normally a mixture of crude oils is sought in the refinery load so that it does not exceed 0.5 mg KOH/g, above this reference, the bottom sections of the distillation units can undergo a severe corrosive process, leading to shorter periods of operational campaign and higher operating costs in addition to problems associated with integrity and safety. Naphthenic acidity is directly linked to the concentration of oxygenated compounds in the crude oil that tend to be concentrated in the heavier fractions, giving instability and odor to the intermediate currents.

The sulfur content in crude oil is another key factor in the naphthenic corrosion phenomena. In crude oils with a sulfur content higher than 2,0 %, a protective layer of iron sulfide (FeS) is formed in the metal surface that is insoluble, avoiding then the attack by naphthenic acids as presented in Figure 9.4. In this sense, refiners processing very low sulfur crude oils with high total acid number (TAN) can face severe issues with naphthenic corrosion in their refining hardware.

Equation (9.1) presents the chemical representation of naphthenic corrosion.

$$Fe + \mathbf{2RCOOH} \rightarrow Fe\left(RCOO\right)_2 + H_2 \tag{9.1}$$

According to the literature, above 400°C, the iron naphthenate (corrosion product) is soluble in hydrocarbons and is attacked by hydrogen sulfide (H_2S) leading to the regeneration of the naphthenic acid, as represented in Equation (9.2).

$$Fe\left(RCOO\right)_2 + H_2S \rightarrow FeS + \mathbf{2RCOOH} \tag{9.2}$$

In crude oil distillation units, the bottom section of the atmospheric column and the vacuum distillation column are the most common regions where naphthenic corrosion is observed while in delayed coking units the phenomenon is observed in the bottom section of the main fractionator's column.

The carbon steel, series 300 and 400 stainless steel, and nickel alloys tend to suffer naphthenic corrosion.

Among the actions to control the naphthenic corrosion in the refining hardware is the blending of crude oils aiming to keep the TAN and sulfur content inside the adequate limits. Other alternatives are the injection of neutralizers or corrosion inhibitors in the processing streams and the selection of materials with higher resistance to naphthenic acid attack.

Refiners processing crudes with high acidity and low sulfur normally apply chromium and molybdenum alloys in the bottom sections of crude oil distillation columns and transfer pipes as well as in residue upgrading units. In extreme conditions, it's possible to consider applying stainless steel 317 L that presents high resistance to naphthenic corrosion, this decision needs to consider the higher capital investment due to the high cost of this material.

As inspection and monitoring strategies, ultrasound and X-ray assay are normally applied to identify localized corrosion and thickness loss as well as sacrifice metals coupons to determine the corrosion rates. Naphthenic acid corrosion is not expected to be a serious issue for the stockpiling assets; however, the knowledge of this corrosion mechanism is important for addressing possible issues related to the pipelines dedicated to fed or withdrawing hydrocarbons for crude oil distillation and delayed coking units.

9.2 CORROSION PROTECTION

Unfortunately, the corrosion process is thermodynamically favorable. It means that the corrosion process can be controlled but not totally stopped, or in other words, the corrosion process will lead the metals to the less energy-intensive state, which is the oxidated compounds; this makes sense if we consider that the production of the metal is a highly intensive energy process.

By understanding the corrosion process, it is possible to develop and design adequate strategies to control the corrosive processes. The corrosion control strategies can be classified, as presented in Figure 9.5.

During the design phase, it is possible to take actions capable of making the assets intrinsically more resistant to the corrosion processes. The most applied strategies to control the corrosion process in the design stage are:

- Adequate material of construction (MOC) selection – Considering the information regarding the process – such as fluid, contaminants concentration, and temperature, – it is possible to carry out an adequate materials selection aiming to ensure adequate resistance to the equipment or tubulation at least to avoid premature failures. The main criteria applied to select the materials for processing equipment are the cost, chemical, mechanical, and physical properties, including availability.

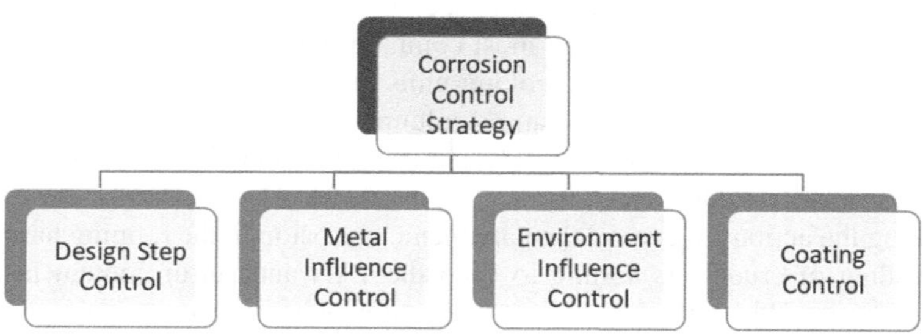

FIGURE 9.5 Classification of corrosion control strategies.

- Materials compatibility – As described earlier, some materials are incompatible when operating in electrochemical contact leading to galvanic corrosion and this fact should be considered during the design stage or during the maintenance of process equipment once it leads to the insertion of failure point or weakness in the process.
- Changes in the environment conditions – In some cases, the corrosion control can be reached by changing process variables which will reduce the corrosivity of the environment where the process equipment is operating. Changes in variables – such as pH, temperature, humidity, or flow rate – can minimize the severity of the corrosion process and enlarge the range of materials that can be applied to the process.
- Components geometry – Normally, a good process equipment design will avoid crevices and sharp corners once these arrangements will lead to contaminants concentration and tension accumulation over the material which expose the process equipment to crevice and stress corrosion processes, respectively. This point is especially relevant for stockpiling assets which concentrate the most part of the tubulations and pipes in the refinery; the design of the tubulations and pipelines should avoid sudden changes in the flow direction to avoid the corrosion–erosion process.

The corrosion control through the metal influence is one of the most applied corrosion control strategies, especially in stockpiling assets. In this strategy, actions are taken in order to raise or reduce the oxidation potential to reach the protection of the metal of the process equipment; this can be reached through cathodic or anodic protection.

The cathodic protection of a metal can be reached through basically two strategies:

1. Impressed current – An external power source is connected to the structure or process equipment and the anode;
2. Sacrificial anode – In this case, pieces of metal with high oxidation potential are installed rather than the process equipment metals aiming to force the equipment shell to be the cathode and forcing the corrosion process to the sacrificial anodes. This strategy is commonly applied to control the corrosion process of the bottom plates of storage tanks, mainly those dedicated to the storage of crude oil.

In the anodic protection strategy, the metal is led to a passivity state reducing then the corrosion rates. This can be reached through the formation of a protective layer over the metal composed of metal oxide. It is important to highlight that this protective layer will reduce significantly the corrosion rates, but it is not capable of stopping the corrosion process.

Due to their characteristics, anodic protection cannot be applied to metals that do not oxidize in the process operation environment and the superficial irregularities over the metal need to be avoided once will lead to failure of the oxide layer, creating the worst scenario of great area of the cathode (passivated metal) and small area of the anode (exposed metal) creating high anodic current intensity which will lead to high corrosion rates. Stainless steel is among the most common materials where anodic corrosion protection is effectively applied.

Environment influence control involves modifying the process environment through the manipulation of process variables, aiming to reduce the corrosion rate and protect the process equipment. By changing variables such as temperature, composition, fluid velocity, and contaminant concentration, it is possible to reduce corrosion rates and extend the operational lifecycle of process equipment.

Among the composition changes, an example is breakdown of passivity in stainless steel under chloride concentration. Reducing the chloride concentration can be an effective strategy to minimize the corrosion process in stainless steel equipment.

The fluid velocity is another variable that can be managed to control the corrosion process. As mentioned earlier, the corrosion–erosion process is related to excessive velocity which removes the protective layer of the metal exposing the base metal to the corrosive process. Besides, the excessive velocity can lead to the cavitation process which will raise the corrosion rates, and lead to the premature failure of process equipment.

Regarding the temperature, it is possible to change the corrosive behavior of a metal by changing the temperature. An example is the change in the galvanic couple behavior of zinc (Zn) and iron (Fe). Under environmental temperature, the zinc is anodic in relation to iron, but when the temperature is raised above 60°C, the zinc becomes cathodic in relation to the iron. Another effect of the temperature is to reduce the oxygen solubility in the fluid which affects the protective layer formation over the metal, influencing the corrosion rates.

The corrosion control through coating consists of using a coating over the metal to be protected aiming to create a barrier between the corrosive environment and the metal.

9.3 THE THREAT OF "DEAD LEGS"

One of the most important challenges to asset management in the stockpiling area is dealing with the "dead leg" tubulations. These tubulations or tubulations sections present none of low-velocity flow rate (stagnant flow), leading to high corrosion rates.

The stagnant flow rate will lead to an adequate environment where microorganisms can develop colonies and proliferate, leading to deposits which will create an adequate scenario for high corrosion rates based on under-deposit corrosion (a type of concentration cell corrosion). Despite the "dead legs" tubulations without flow rate, these tubulation sections are still connected to the process, leading to a severe risk of hydrocarbon contention loss, which can lead to severe process safety accidents.

The "dead legs" threat is a significant concern in the downstream industry both on-site and off-site (stockpiling assets) of the crude oil refineries. The APIs 574 and 581 deal with "dead legs" issues and establish the orientations to minimize the risk of contention loss due to the "dead legs" corrosion.

Ideally, the "dead legs" issues should be considered in the design step of any industrial facility avoiding tubulation sections with stagnant or low-frequency flow rates, but this is very difficult to reach mainly considering long tubulation arrangements like the stockpiling area of crude oil refineries. Despite this fact, it is important to consider this issue during the design step and management of change in stockpiling assets, it is rare for refinery personnel to conduct process changes in stockpiling assets by deactivating tubulations and pipeline sections, simply blocking the flow rate to the tubulation using a valve or a mechanic device as line blinds. This practice creates "dead legs" and creates a new threat to the process. The "dead legs" can be classified into three main categories:

- Permanent "dead legs" – Tubulation sections where flow rate is not accepted, in this case, it's recommended to proceed with a management of change to eliminate this tubulation section, eliminating the process safety risk;
- Temporary "dead legs" – Tubulation sections temporally without flow rate; in this case, an adequate flushing procedure should be conducted to minimize the corrosion risk or involve the equipment inspection team to define the maximum period of time that this section could be kept under this condition;
- Operational "dead legs" – These assets are normally used by the operation teams under specific conditions like shutdown and start-up of processing units or storage tanks; this can be very common in the operation of stockpiling assets in crude oil refineries. In these cases, the best management route is to establish in accordance with the equipment inspection team, the adequate frequency and inspection technique to monitor the corrosion rate and integrity of the tubulation sections.

To avoid this, the management of change in stockpiling assets should always be conducted by a multidisciplinary team led by a process engineer with expertise in the asset as well as an inspection engineer aiming to avoid the superficial analysis which can raise the risk of stockpiling assets. One of the key reliability tasks that should be conducted by the stockpiling manager is a deep analysis and identification of "dead legs" in the stockpiling assets aiming to allow an adequate management of these points, as suggested in Figure 9.6.

The first step of the "dead legs" management strategy is to identify the "dead legs" tubulations followed by the classification as described above. The "dead legs" classified as permanent should be eliminated (if possible) through adequate management of the change process aiming to eliminate the process safety risk associated with this threat while the other "dead legs" registered and communicated to the equipment inspection and operation teams, for the cases where the flushing procedure is possible. This procedure can be defined and communicated to the stockpiling team aiming to ensure that the flushing procedure will be carried out under the defined frequency, minimizing the process safety risks associated.

For the "dead legs" where the flushing process is not possible, the equipment inspection team should define an adequate frequency of inspection with adequate technique aiming to determine the corrosion rate and the remaining life of the tubulation as well as define the best strategy to repair the low thickness points.

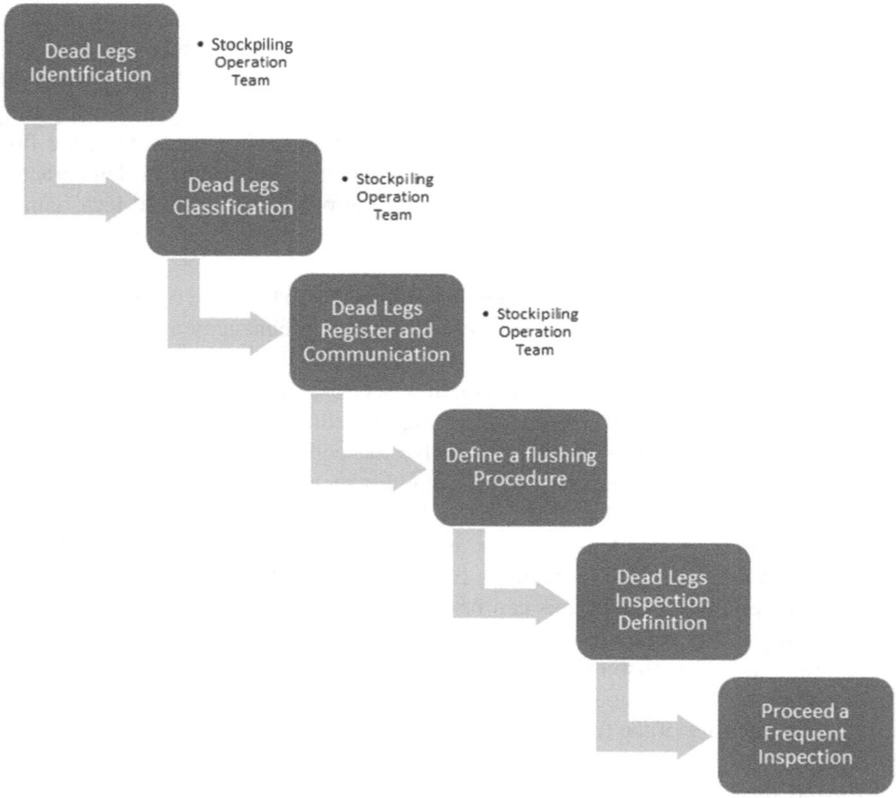

FIGURE 9.6 Proposed strategy to manage "dead legs" in stockpiling assets.

9.4 CORROSION PROCESS IN STORAGE TANKS

Another key question regarding the asset integrity of stockpiling assets is related to the corrosion process in storage tanks. Throughout the history of crude oil refining history, it is not uncommon to hear about contention loss events and process safety accidents caused by hydrocarbon leakages from storage tanks.

The most common scenario is the contention loss from hydrocarbon storage tanks due to failure of the bottom provoked by the corrosion process. However, roof corrosion is another important concern to refiners once, as it can allow the contamination of the inventory with rainwater, leading to quality issues of the derivatives. Additionally, roof corrosion poses safety risks for operators who need to access the tank roof for routine tasks such as verifying tank gauging.

Regarding the roof corrosion process, the main corrosion process present is related to the presence of acid compounds in the vapor space between the liquid surface and the roof, as presented in Figure 9.7.

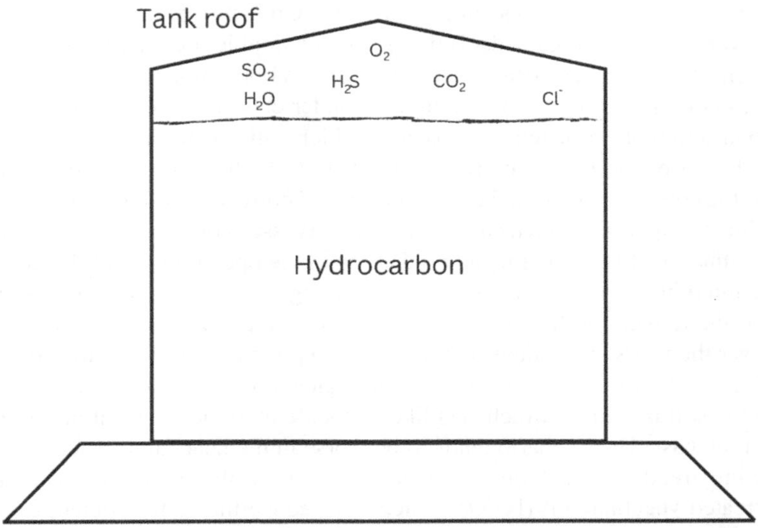

FIGURE 9.7 Corrosion phenomena of hydrocarbon storage tank roof.

The composition of the vapor space relies on the stored hydrocarbon, but it is expected that heavy and high contaminants concentration like fuel oil and heavy sulfur diesel tends to present high sulfur compound concentrations which will raise the corrosion rates of the roof. As presented in Figure 9.6, the most common corrosive compounds present in the vapor space of hydrocarbon storage tanks are CO_2, SO_2, H_2S, Cl^-, O_2, and H_2O.

The internal environment of the storage tanks favors the corrosion process due to the high relative humidity and temperature (which can reach values above 60°C). These conditions plus the presence of acid compounds will lead to acid condensation over the roof sheet reducing the pH to values under 3 which will lead to higher localized corrosion rates.

Unfortunately, the roof corrosion process is responsible for a significant reduction in the operating life of storage tanks. Over the years, the refiners developed and tested some corrosion control techniques such as roof coating, special paints, and material change aiming to control the corrosion process. The most applied strategy by the refiners for the storage tanks which presents chronic issues regarding roof corrosion is to change the roof sheet material from carbon to stainless steel which presents higher corrosion resistance.

Another alternative to control the roof corrosion process in storage tanks is to use volatile corrosion inhibitors (VCI). These compounds adsorb on the surface of the roof sheet suppressing the corrosion reactions, typical compounds applied as VCI are organic amines and carboxylates.

Regarding the bottom plate corrosion, this is considered the higher risk corrosion process once will be led to an immediate contention loss with significant risk of fire of the storage tank. Nowadays, aiming to mitigate this process safety risk, most part of the hydrocarbon storage tanks is designed with double bottom plates.

The most common corrosion mechanism in the bottom plates of storage tanks is localized corrosion which can be pitting or under-deposit corrosion associated with microbial action of sulfur-reducing bacteria which will lead to sulfuric acid formation which decreases the pH of the region and lead to higher localized corrosion rates. Further the corrosive process, the microbiological corrosion in storage tanks can lead to biofilm precipitation which can provoke quality issues of the derivatives like diesel and gasoline like filter plugging and reduction of the operating life of the tank.

The microbiological corrosion issues of storage tanks became a significant concern for the refiners in the last years due to the change in the derivatives specifications over the years. Nowadays, the necessity to produce very low sulfur derivatives like diesel and gasoline raises the microbiological corrosion risks due to the reduction in the sulfur content which acts like a biocide in higher concentrations as practiced in the past. This scenario tends to be worse in the near future once there exists a growing trend to use biomass in coprocessing with fossil raw materials like hydrotreated vegetable oil (HVO), which will reduce the sulfur content even more and raise the water retention of the final blending.

One of the most effective ways to minimize the corrosion process in the bottom plates of hydrocarbon storage tanks is to minimize the water content in the bottom of the tanks. The first step here is to understand that the stockpiling assets are designed to deal only with the emulsified water which will be separated by decantation in the tanks and then drained; for this reason, it is fundamental to ensure that the storage tanks rely on adequate and reliable draining systems. This is especially true for storage tanks dedicated to the storage of diesel, gasoline, and naphtha.

The storage tanks in crude oil refineries are normally constructed over a concrete foundation aiming to avoid the contact of the bottom plate with the soil as well as to ensure rigidity to withstand the bottom plates, mainly for high-diameter tanks. Despite the strategy to isolate the bottom plate from the soil, moisture accumulation may occur between the bottom plate and the concrete foundation. This issue can be minimized through an adequate design that uses deflecting plates to seal the gap between the bottom plate and concrete pad, avoiding water accumulation in this region which fatally will lead to corrosion.

A key question related to the integrity of the storage tanks is an adequate evaluation of the conditions of the bottom plates during maintenance shutdowns. The visual inspection of the equipment can help but is not sufficient to ensure adequate identification of corroded and fragility points which will reduce the operating lifecycle of the tank. To improve the accuracy of the tank inspection during the maintenance window the inspection team used to apply nondestructive testing or analysis aiming to identify corroded and low thickness points over the bottom plates, this information is applied to determine the local where the plate should be changed or reinforced, defining the maintenance extent of the storage tank.

One of the most applied nondestructive tests to evaluate the condition of tank floors is the magnetic flux leakage (MFL) is described in Figure 9.8.

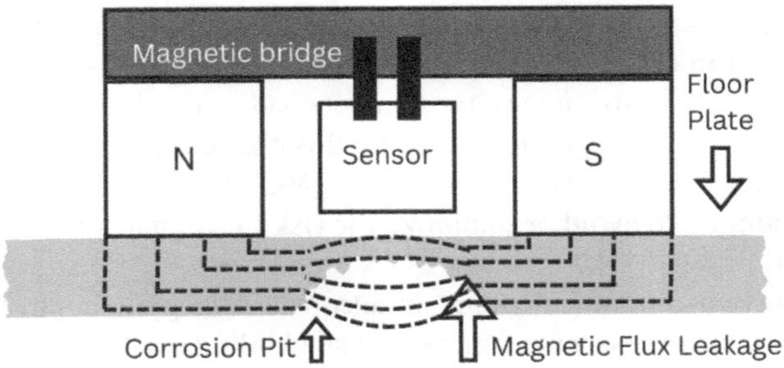

FIGURE 9.8 Basic principle of the magnetic flux leakage (MFL) testing.

The basic principle behind the MFL testing is to produce a magnetic flux through two poles which is reduced when the flux finds a defect or a low-thickness region over the storage tank floor. Using the report produced by the MFL testing the equipment inspection team will define the maintenance scope of the storage tank, which can vary from spot repairs to the total substitution of the floor plates of the tank.

For crude oil storage tanks is relatively common to find significant thickness loss of the floor plates during the maintenance windows considering the characteristics of the fluid (high salt and water concentration) which normally leads the inspection team to recommend the total substitution of the floor plates. For crude oil storage tanks, another key inspection factor is the condition of the sacrifice anodes which are installed in the floor plates to ensure protection through cathodic protection method. If the sacrifice anodes are found in poor condition, it's fundamental to substitute the pieces and the tanks should never be in operation without these devices, and under risk of accelerating the corrosion process, which can lead to contention loss and process safety accidents.

The corrosion of above ground storage tanks' bottoms is one of the main concerns to the stockpiling operation team and asset integrity personnel. The main objective in managing the corrosion rate in the bottom plates is to ensure that the storage tank will reach the next scheduled inspection without loss of containment of the stored hydrocarbons.

The key factors to define the strategy of corrosion control in the bottom of the storage tanks are:

- Foundation type
- Bottom type (double or single)
- Tank pad

The use of double bottom is a widespread practice in the refining industry aiming to control the leakage risks associated with hydrocarbon contention losses due to the corrosion process.

Tank pad selection is another relevant topic that can help in the corrosion control of the storage tank bottom, contamination of the tank pad with corrosive electrolytes such as water will raise the corrosion rates. The most common corrosion pad applied in crude oil refineries is concrete due to its simplicity and effectiveness, despite these advantages, the concrete pad can make it hard to monitor and control the corrosion process. To overcome this challenge, some refiners adopt the strategy to inject volatile corrosion inhibitors (VCI) between the bottom plates and the tank pad aiming to avoid or minimize the risk to reach a corrosive environment in this region.

As previously mentioned, defining an adequate strategy to control the corrosion process is crucial in the asset management of stockpiling assets, especially the above-ground storage tanks. The most common strategies applied to control the corrosion process in the bottom of storage tanks are the cathodic protection method and the VCI application as previously mentioned.

The cathodic protection is applied in the storage tank bottom using sacrificial anodes, impressed current, or both. As previously described, the protection is achieved by forcing the bottom plates of the tank to act as the cathode in the electrochemical cell, forcing the corrosion process to the sacrificial anode. The API 651 presents the details for the design of cathodic protection for above-ground storage tanks.

Although it may seem obvious, the cathodic protection system of storage tanks should be considered a critical system to the asset integrity of the stockpiling assets of a crude oil refinery and demands specific maintenance plans.

As mentioned above, it is fundamental to build a good relationship and partnership between the stockpiling operation, maintenance, and inspection teams aiming to build adequate asset integrity policies for the stockpiling operations. Figure 9.9 presents a proposed model to ensure an adequate asset integrity policy for stockpiling assets.

It is important to understand that the stockpiling assets normally do not have defined maintenance shutdowns like the processing units and the failure of stockpiling assets tends to have a more delayed impact on the refinery production planning than the processing units assets leading to a trend of stockpiling assets losing priority in the daily maintenance schedule. This scenario can lead to a quick degradation of the integrity of these assets, mainly the storage tanks.

As presented in Figure 9.9, the synergy between the stockpiling operation, equipment inspection, and maintenance teams is fundamental to ensuring the integrity of the stockpiling assets.

The equipment inspection should be based on risk analysis considering the inspection's history of the equipment, mainly considering fabrication defects, contamination with corrosive agents from the process (chloride, ammonium salts, etc.), and the extension of the last maintenance and inspection. Postponing a scheduled inspection of a storage tank, for example, should be avoided and if necessary, needs to be carried out after a deep analysis of a multidisciplinary team aiming to ensure that all factors and risks were considered and mitigated.

As described in this chapter, the adequate management of stockpiling assets is a hard and challenging task for the refiners. Considering the impact of the failure and unavailability of these assets over the refining business, especially the storage tanks,

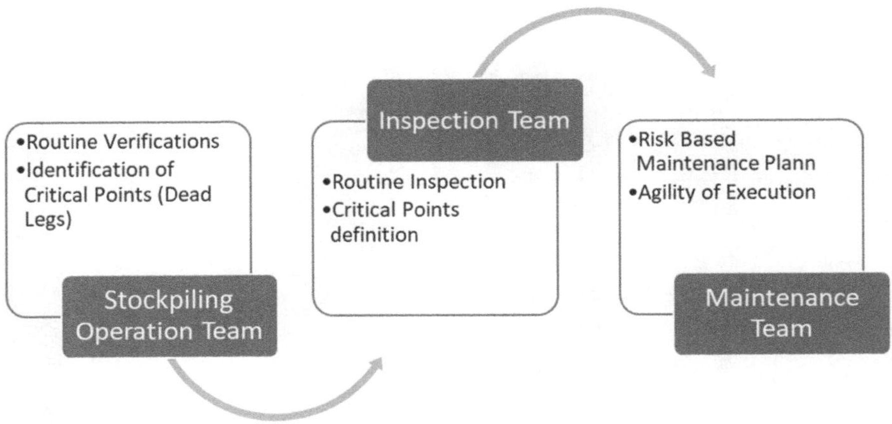

FIGURE 9.9 Asset integrity policy model for stockpiling assets.

an adequate management process of stockpiling assets should be a priority for the refinery management teams and needs to be one of the pillars of the strategic planning of the crude oil refineries.

Considering their characteristics, capital costs involved in the maintenance, and impact on the refinery production planning, the storage tanks demand the creation of a special group that will focus permanently on scheduling, planning, executing, and monitoring the maintenance of storage tanks. This group should be composed of a multidisciplinary team involving at least the following areas: stockpiling operation, static equipment maintenance, equipment inspection, engineering, health, safety, and environment (HSE), remaining maintenance disciplines, production planning, optimization, etc. Figure 9.10 presents a proposal of composition and role definition for the storage tanks maintenance team of a typical crude oil refinery.

The composition proposed in Figure 9.10 is only a suggestion that can be customized according to the characteristics of each refinery. It is fundamental to adequately define the storage tanks maintenance scheduling aiming to avoid the accumulation of storage tanks under maintenance due to the impact of the refinery production planning and the difficulties in managing the high amount of complexity and risk maintenance tasks in parallel.

In some cases, the accumulation of storage tanks under end life and the impact of this unavailability over the refinery results can be the driving force to postpone some maintenance of this equipment. As quoted above, this decision needs to be deeply studied and be adequately supported by the inspection team aiming to avoid new accumulation of storage tank maintenance in the future and, in the worst scenario, process safety accidents caused by asset integrity failures. The decision to postpone an internal inspection of a hydrocarbon storage tank should be only considered after the execution of a nondestructive analysis capable of ensuring or at least reducing the failure probability under operation.

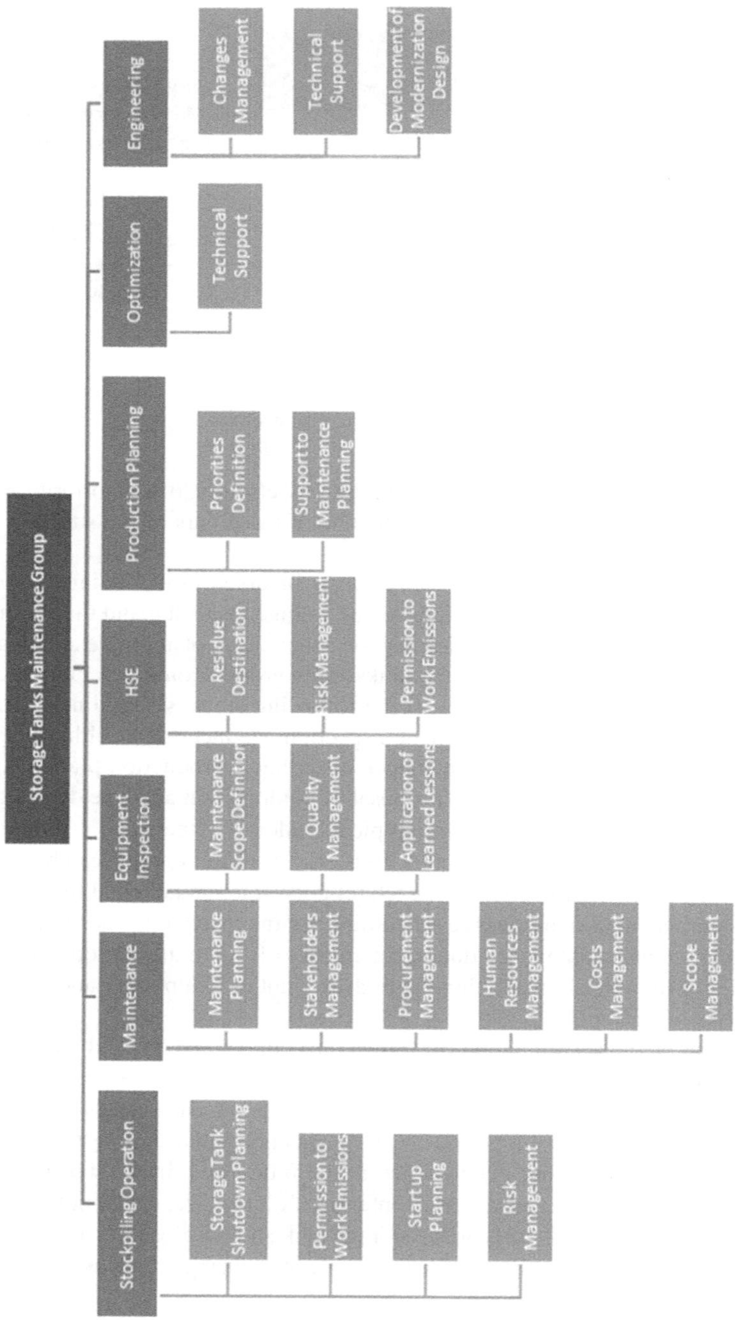

FIGURE 9.10 Proposed composition and roles of the storage tanks maintenance group of a typical crude oil refinery.

9.5 THE THREAT OF FLUID HAMMER IN STOCKPILING ASSETS

Another great threat to the integrity of stockpiling assets is the fluid hammer phenomenon. The fluid hammer is defined as a shock wave produced by a sudden pressure variation inside pipes resulting from flow variations imposed on the flow of liquids. The most common causes or contributors for fluid hammer occurrence are:

- Sudden opening and closing of valves
- Failure of protective devices like steam traps
- Sudden stoppage of dynamics equipment.

These flow characteristics can lead to hydraulic transients where a liquid mass is chocked with the pipeline and accessories causing vibration, noise, and potential failures of the pipes. The following operational systems demand special attention regarding the risks of fluid hammer occurrence:

- Low-pressure steam and condensate pipelines (higher probability of condensation)
- Hydrocarbon pipelines with the possibility of two-phase flow (unit load lines)
- Steam pipelines with poor thermal insulation (overload of the steam trap system)
- Pipes subject to sudden pressure variations (e.g., water supply)
- Emergency water supply pipelines

It is fundamental to consider that the fluid hammer is not a normal operating condition. The occurrence of this phenomenon is related to failures of design, maintenance, and process adjustment under no circumstances can the operating personnel accept to coexist with the fluid hammer occurrence, in some cases, the hydraulic transient can lead to severe contention loss and process safety accidents.

BIBLIOGRAPHY

Chang, J. I.; Lin, C.C. *A Study of Storage Tanks Accidents*, Journal of Loss Prevention in the Process Industries, Elsevier, 2006.

Control of Corrosion Under Thermal Insulation and Fireproofing Materials, NACE Standard Practice SP0198-2016, NACE International, 2016.

Corrosion Under Insulation and Fireproofing, American Petroleum Institute, Recommended Practice API 583, 2014.

Ramanathan, L.V. *Corrosion and his Control*. 1st ed. Hemus Press, 1978.

Robinson, P.R.; Hsu, C.S. *Handbook of Petroleum Technology*. 1st ed. Springer, 2017.

Silva, M. W. Personal Notes.

Silva, M. W.; Clark, J. *Flexible Upgrading of Heavy Feedstocks*. PTQ Magazine, 2021.

Sutton, I. *Plant Design and Operations*. 2nd ed. Gulf Professional Publishing, 2017.

10 Managing the Interface between Refinery Stockpiling Operations and Middle Stream Companies

A crude oil refinery can be compared with a life organism once each part of the refining process has an impact on the whole process. The refinery personnel and especially the management team should understand this fact and this needs to be considered in the strategic planning of the refinery.

The crude oil refinery is supplied with crude oil and their products or derivatives are withdrawn through an intricated pipeline network. It is clear that the adequate operation of these pipelines is fundamental to a sustainable operation of the crude oil refinery which can shut down due to the lack of crude oil or due to exhaustion of derivatives storage space in case of pipeline unavailability; this scenario reinforces the relevance of adequate design of the tank farm of a crude oil refinery aiming to ensure adequate time to absorb process variations and adverse scenarios with minimum impact over the refinery production planning, process safety, economic results, and market supply.

To support the logistics operation of the crude oil supply to the refinery and the derivatives distribution to the final consumers, there are companies specialized in hydrocarbon transfer operations called midstream companies. Considering the high synergy between the midstream companies and the crude oil refineries, adequate management of this interface is another critical task for the refinery management team; this is especially true for the stockpiling and production planning teams.

Normally, the pipeline network does not attend exclusively to a specific crude oil refinery, but many refineries can be connected leading to adequate planning and scheduling to avoid market shortages, impacts over the refinery production planning, and long idleness period of the pipelines which means higher operating costs.

10.1 MANAGING THE THREAT OF THE MIDSTREAM COMPANIES TO THE REFINERY

As introduced above, the results of the crude oil refinery are totally dependent on the adequate operation of midstream companies aiming to ensure an adequate supply of crude oil as well as the withdrawal of crude oil derivatives before reaching the installed storage capacity in the tank farms of the refinery.

DOI: 10.1201/9781003472216-10

Considering the relevance of this synergy to the whole downstream business, the stockpiling should be able to identify significant threats in the midstream operations which can produce impacts over the refinery, mainly related to asset management. As the refinery stockpiling area, the midstream companies have a significant number of pumps and kilometers of pipelines to manage aiming to ensure the crude oil supply and derivatives withdraw from the refinery.

Despite being a responsibility of the midstream company, the stockpiling manager needs com stablish adequate management processes in order to ensure adequate information flow capable to supply in detail the main threats to the midstream operations and consequently, over the refinery results.

The stockpiling manager should make adequate proximity management of the midstream management personnel to ensure this information flux and make periodic meetings aiming to monitor the progress of the maintenance process of critical equipment such as pumps, valves, and pipelines. If possible, the main equipment of the midstream companies which has a direct impact on the refinery operations should be monitored in the reliability panel of the crude oil refinery.

The main goal of this approach is to detect, in adequate time, the threats related to midstream processes and assets which can cause significant impacts on the refinery operations. These threats can be adequately informed to the refinery management team as part of the support decision package and should be considered in the production planning of the refinery.

As previously mentioned, despite not being under the direct management of the refinery, the quality of asset management of midstream companies needs to be monitored by the stockpiling manager through adequate management processes. Good practice is monitoring the reliability and availability of midstream assets like pumps and pipelines in the critical equipment panel of the refinery. This panel should concentrate on all equipment capable of causing process safety issues and shutdown of processing units. In addition, frequent meetings to monitor this panel should be carried out defining action plans to mitigate the risks and defining actions and tasks capable of reducing the time of unavailability of these assets.

10.2 THE MUTUAL OPERATIONS BETWEEN MIDSTREAM AND CRUDE OIL REFINERIES

Considering the relevance of the mutual operations between the midstream and refining assets, it's necessary to ensure that the operations are covered with adequate procedures and that all process safety risks are charted and monitored.

The main document that drives the operations between midstream and the refinery or stockpiling refinery area is the mutual operation procedure. This document is developed involving all stakeholders and defines the basic and minimum safety procedures aiming to minimize the process safety risks and operating failures like tank overfill and hydrocarbon contention loss during transfer operations in pipelines.

The main operation mutual procedure needs to define the operational hierarchy over the pipeline and define, clearly, the responsibilities over each operation. Furthermore, the document needs to establish the operating and safety limits of the pipeline as well as the critical process control variables and the time interval for an adequate monitoring of these variables.

Another key factor that should be defined by the mutual operation procedure is the communication protocol between the midstream and downstream agents aiming to ensure effective and clear communication, especially in emergency scenarios. The document should define the actions that need to be taken in abnormal situations, mainly related to process safety issues.

Furthermore, the mutual operation procedure should contain a list of equipment and instruments that are considered critical for pipeline operation as well as the definition of how the operating issues and abnormalities will be registered and managed allowing the companies to learn from the operational occurrences and near accidents.

Operations involving the refinery stockpiling personnel and the midstream operations teams should be adequately managed, especially during the transitory step. The transitory state or regime is the time interval between the transfer starts and the receiving tank indicating level variation.

During this period, it is necessary to a rigorous control and monitoring of the critical variables like pressure and flow rate in the pipeline. When the receiving tank starts to receive the transfer under a near-constant flow rate, we can consider that the transfer operation is in under steady state. During the steady state, the transfer is constantly monitored, and the transfer should be stopped when the pipeline flow rate is higher than 2 to 5% of the flow rate in the metering station at the interval of 15 minutes, this limit is established aiming to prevent leakage along the pipeline.

At the end of the transfer operation, it is possible to identify variations of the process variables like level, flow rate, and pressure. This period is called the accommodation step and it is important to ensure that the pipeline valves are adequately sealed aiming to avoid undesired movements as well as the pipeline should be kept under expansion avoiding overpressure in the system.

Another fundamental document that is normally applied to help the adequate management of the operations involving interfaces between the refinery stockpiling area and the midstream companies is the transfer authorization; this document should be emitted by the receiver agent and should contain, at least, the following information:

- Identification of the sending tank
- Identification of the receiving tank
- The product which will be transferred (gasoline, diesel, etc.)
- The available space in the receiving tank
- The volume which will be transferred
- The operating flow rate
- Confirmation that the necessary alignment was concluded ensuring that the receiving tank is ready to receive

As previously mentioned, the transfer authorization aims to avoid process safety accidents involving tank overfill and overpressure in the pipelines which can lead to severe emergency scenarios.

10.3 THE INTERFACE BETWEEN THE REFINING PRODUCTION PLANNING AND SCHEDULING AND THE STOCKPILING AREA

Considering the high capital investment involved in the implementation of crude oil refineries as well as the high fixed costs involved in sustaining the refinery under operation, the refining asset should be optimized aiming to ensure maximum profitability, or in common terms of the downstream industry, high refining margins.

Refiners' profitability is directly proportional to their capacity to add value to the processed crude oil, aiming to maximize the production of high-added value streams and derivatives. Equation (10.1) presents a simplified concept of the liquid refining margin.

$$Liquid\ Refining\ Margin = \sum_{i}^{n} \left(Di\ x\ Vi\right) - Pc - \left(Fc + Vc\right) \tag{10.1}$$

The first term in Equation (10.1) corresponds to the obtained revenue through the commercialization of the crude oil derivatives, represented by the sum of the product between the derivative market value and the volume or weight commercialized. As aforementioned, the profitability or refining margin is directly proportional to the refinery capacity to add value to the processed crude slate, the maximization of higher added value derivatives leads to the maximization of the first term in Equation (10.1).

The players capable of maximizing the yield of higher added value crude oil derivatives in the refining hardware reach a highlighted competitive advantage in the downstream market.

Maximizing the refining margins is one of the main roles of the production scheduling and planning of a crude oil refinery.

The maximization of high-added value derivatives is directly related to the adopted refining scheme, Figure 10.1 presents a basic process flow diagram of a typical refinery operating with the "Topping" refining configuration, considered the simplest and with the lower complexity of the available refining schemes.

Nowadays, this refining scheme is less practicable once the purely physical processes difficult to achieve an economically attractive level of crude oil conversion; furthermore, the produced derivatives have high contaminants content, especially sulfur and nitrogen, making prohibitive his commercialization as final derivatives without infringing the current environmental regulations, in this case, the derivatives can be commercialized as intermediate streams that have relatively low added value. The quantity of asphalt and fuel oil which are products with low added value and restricted market produced in a Topping refinery makes his operation uncompetitive economically. In these cases, profitability is normally reached only under the processing of specific crude oils (lighter and with lower levels of contaminants), which have high acquisition costs and restricted market supply, leading to a competitive loss.

The addition of higher capacity for treating and conversion, especially of bottom barrel streams, makes the crude oil refining scheme more complex; however, raises the yield of high-added value derivatives, contributing positively to improving the

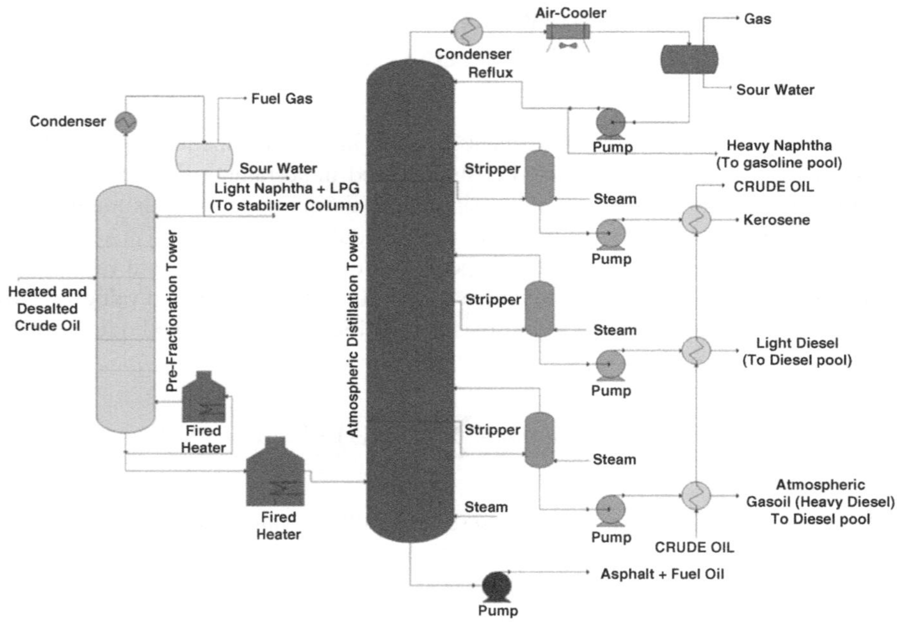

FIGURE 10.1 Typical process configuration to a refinery with "topping" refining configuration.

refining margin, according to Equation (10.1). Figure 10.2 shows a typical process flow scheme for a refinery operating under the coking/hydrocracking refining configuration. It is important to highlight that the rise of refinery complexity also tends to raise operational costs.

In the coking/hydrocracking refining configuration, the production of fuel oil and asphalt is reduced to the minimum needed to meet the available market consumer, and delayed coking and hydrocracking units raise the production of higher added value derivatives such as naphtha, diesel, and kerosene that allow a significant improvement in the refiner profitability.

Even in complex refineries, a common strategy to maximize the liquid refining margin is to add to the high-added value derivatives the maximum quantity of streams with low-added value to compose the final derivative. However, it is necessary to meet some restrictions, as presented in Figure 10.3.

The hydrocarbon mixture that will compose the final derivative needs to respect some restrictions such as the available component stocks, market demand, the refining hardware limits, keep the prices attractive, and mainly meet the quality and environmental regulations. The region limited by these restrictions is called "blending space" and respecting these restrictions, is possible to blend the low-added value streams to compose the final derivatives.

In Equation (10.1), the term *Pc* corresponds to the acquisition cost of crude oil, the refiners normally do not have control over the market, and the acquisition cost of petroleum is controlled by the geopolitical scenario and the international market. The cost reduction in the crude oil acquisition can be achieved through the

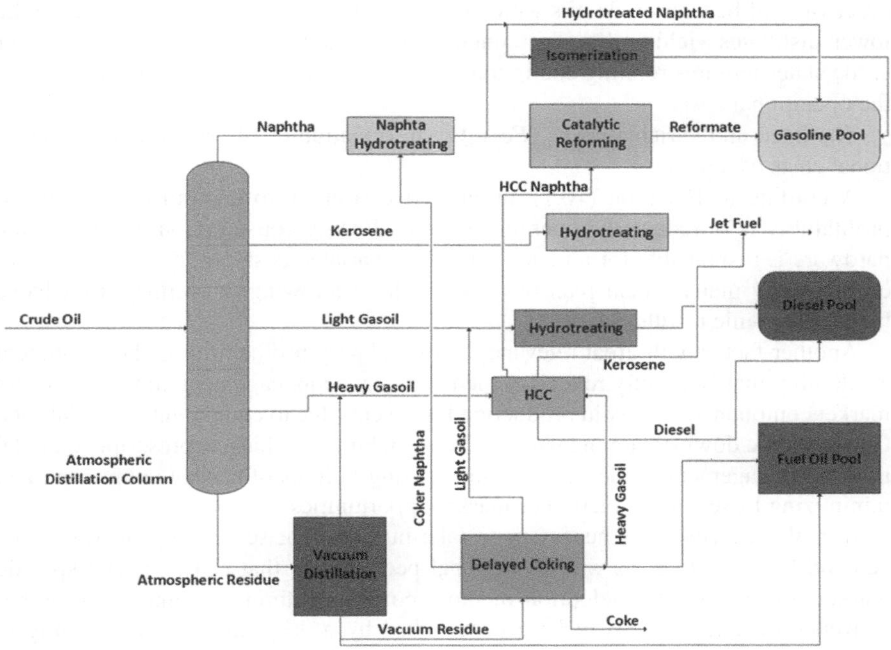

FIGURE 10.2 Process arrangement for a typical refinery operating under coking/hydrocracking configuration.

FIGURE 10.3 Schematic representation of the "blending space" to crude oil derivatives.

processing of heavier crude oils which have relatively low cost; however, due to the lower distillates yield and higher contaminants content, the processing of heavier crude slates requires refining configurations with higher conversion capacity, raising the operational costs.

The fixed and variable costs, $(Fc + Vc)$ in Equation (10.1), represent the operational costs of refiners.

According to Equation (10.1), an attractive strategy to maximize the refiners' profitability is to reduce the operational costs. Energy consumption in the refining hardware is responsible for a major part of the variable costs; in this sense, an adequate energy management plan has a key role in allowing the refiners to achieve higher economic results.

Another factor with great relevance to the refinery profitability is the equipment availability that is directly related to the maintenance management aiming to ensure market compliance and avoid production loss events due to equipment unavailability. Currently, the downstream industry has been looking to achieve a prescriptive model in asset management in the sense of anticipating failures of critical equipment, and minimizing losses of production or market opportunities.

In real scenarios, it is necessary to take into account seasonality factors in the demand by derivatives as well as regional peculiarities that can represent specific markets to refiners. The adoption of more complex refining schemes with higher conversion capacity needs to be accomplished by a deep market study aiming to ensure a constant demand by higher added value derivatives once the capital and operational costs is significantly higher; in other words, the revenue expected by the derivatives commercialization need to justify the raise in the operational costs and improve the refining margin, according to Equation (10.1).

Considering these concepts of refinery economics, the role of the planning and scheduling production team is to optimize the refining asset aiming to maximize the refining margins. Figure 10.4 presents an overview of the planning and scheduling of production of a crude oil refinery.

The first step of the production planning process is to ensure the availability and supply of the best crude oil considering the characteristics of the refining asset. This point is fundamental, as previously explained, to ensure the maximum yield of higher added value derivatives such as jet fuel, diesel, and gasoline to the detriment of bottom barrel streams which will compose the fuel oil pool of the refinery.

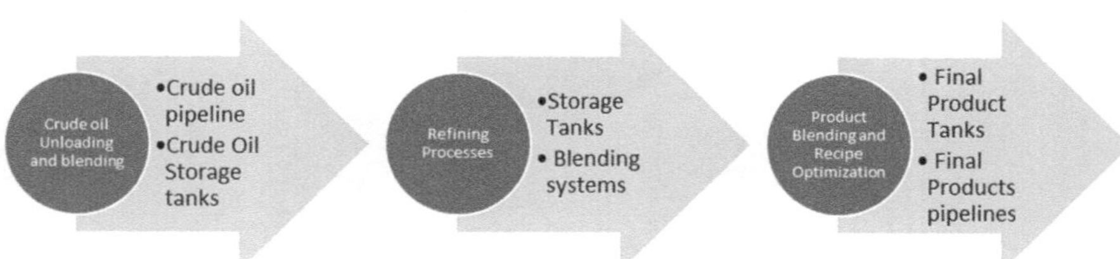

FIGURE 10.4 The systems of a typical crude oil refinery involved in the production planning process.

The second step is – based on the yield profile of the crude oil in the refining asset – to produce the recipes which will be followed by the stockpiling personnel in the blending systems as presented in detail in Chapter 5. The main objective here is to maximize the addition of lower-added value streams in the final blending, respecting the restrictions described in Figure 10.3.

The final step of the production planning process is to optimize and adequately schedule the operation of the final product pipelines aiming to ensure that the refinery will not be forced to reduce the throughput due to lack of storage space in the final product tanks.

Figure 10.4 indicates the stricter relationship between the stockpiling area and the success of the optimization process of the production planning of the refinery. The availability of storage tanks and blending systems is essential to allow the refinery to be able to follow the orientations of the production planning and scheduling team. When the health of the asset management of the transfer and stockpiling area is poor, it is ensured that the refinery will lose money and competitiveness in the market.

Another key rule to keep adequate asset integrity in the refinery is to ensure that the feed tanks of the processing units will be drained during the planned maintenance shutdown of the processing units. It is common event of efficiency loss of pre-heating battery of heat exchangers from hydrotreating units due to the drag of poor quality derivatives trapped in the bottom of the feed tanks; this can occur when the draining routine of the tank is not respected and there is an accumulation of water and hydrocarbons which allow the action of sulfate reduction bacteria leading to the accumulation of solids and organic materials which can accumulate in the heat exchangers of processing units. To avoid this, it is fundamental to keep the operating routine of draining the bottom of the feed tanks as well as verify the conditions of the bottom of the tanks during the planned maintenance of the processing units, cleaning the tanks if necessary.

Considering the strong dependence between the production planning and stockpiling teams, it is necessary to take management actions to ensure adequate integration between the teams. This integration is fundamental to ensure that the teams will be adequately informed of the main priorities and then take actions to ensure the availability of assets to ensure compliance with the production planning.

Nowadays, the refiners apply modern and complex software capable of integrating information and restrictions of the refining hardware aiming to produce the best production planning for the current status of the refining asset.

BIBLIOGRAPHY

Fahim, M.A.; Al-Sahhaf, T.A.; Elkilani, A.S. *Fundamentals of Petroleum Refining*. 1st ed. Elsevier Press, 2010.

Gary, J. H.; Handwerk, G. E. *Petroleum Refining – Technology and Economics*. 4th ed. Marcel Dekker., 2001.

Robinson, P.R.; Hsu, C.S. *Handbook of Petroleum Technology*. 1st ed. Springer, 2017.

Silva, M.W. Personal Notes.

11 Process Safety and Asset Management in LPG and Liquefied Gas Storage Farm

The liquefied gas storage farm concentrates the major process safety risks in the stockpiling area of a typical refinery. The great amount of energy stored in the LPG, propylene, and butane spheres makes an eventual leakage and fire an enormous process safety risk with great potential for destruction.

The liquefied gases are stored in spheric pressure vessels due to their high volatility, which demands reduction in the heat transfer to the storage vessel to keep the hydrocarbon in the liquid phase. Among the available options, the sphere form presents a lower ratio between surface area per volume, which will minimize the heat flux. This will be reflected in less capital spending with vessel insulation. Furthermore, the liquefied gases demand storage under relatively high pressure, and the spheric vessels ensure uniform tension along the vessel, minimizing the process safety risks associated with failures caused by tension concentrations in the storage vessel materials.

Considering the relevance of these assets to the stockpiling area as well as to the whole refining business, we decided to dedicate an exclusive chapter in this book to this topic, aiming to highlight the risks associated with the operation of LPG spheres and how to manage this in order to minimize the risks of process safety events or at least reduce those impacts.

11.1 LIQUEFIED PETROLEUM GASES (LPGs)

The liquefied petroleum gas (LPG) is normally produced from crude oil in the refineries in processing units like crude oil distillation, fluid catalytic cracking units (FCCU), and delayed coking. It's possible to reach LPG production in other processing units like naphtha catalytic reforming, but in small quantities. Another common production route of LPG is the condensation from natural gas in natural gas processing plants, which is normally not associated with the crude oil refinery operations.

The main LPG producer in a crude oil refinery is the FCC unit, of course, when this processing unit is available in the refining scheme. In a typical FCC unit, the volumetric yield of LPG is around 20%. Due to its characteristics, LPG is normally applied as cooking fuel in residences or for heating according to the consumer market.

LPG is normally composed of paraffinic and olefinic hydrocarbons in the range of C3 to C4 and is applied as domestic and transportation fuel in specific cases.

DOI: 10.1201/9781003472216-11

The LPG can contain low quantities of light and heavy hydrocarbons (C2 and C5); however, the concentration of these compounds needs to be minimized, aiming to not prejudice the quality requirements. The concentration of light hydrocarbons is controlled through the Reid vapor pressure (RVP), which is determined by the LPG heating at 37.8°C. The light content is controlled for security reasons, aiming to keep the LPG volatility under safe values to allow storage and handling. The RVP of commercial LPG is controlled below 1430 kPa. Once LPG is normally burned in closed environments, the control of burning residue is one of the most important quality requirements of this derivative. The heavy content is controlled through weathering test that evaluates the difficulty in the vaporization of LPG, measuring in an indirect way the content of C5+ in the mixture, which is normally defined by the boiling temperature of 95% in volume of the mixture under atmospheric pressure and is normally controlled below 2.0°C.

The other quality requirements of LPG are the sulfur content and corrosivity. The sulfur content is controlled below 140 ppm, and this parameter, as well as corrosivity, is managed through desulfurization processes such as regenerative caustic treatment.

A refinery LPG pool is normally composed of the LPG produced in FCC units that are the main producers of this derivative, delayed coking units, catalytic reforming, and atmospheric distillation, and, in some cases, LPG from natural gas treating plants. The LPG from FCC can be submitted to additional separation processes to recover the propylene and direct it to the market as a petrochemical intermediate due to its higher added value when compared with the LPG.

Some refiners can apply the strategy to split the C3 and C4 streams from the LPG production, aiming to allow the C4 addition to the gasoline pool as a strategy to maximize the added value to the C4 stream and improve the octane index of the final gasoline.

11.2 PROPYLENE

Propylene is one of the most important petrochemical intermediates nowadays, being the second largest consumed in the world, behind ethylene. Propylene can be applied as an intermediate to the production of some fundamental products, for example:

- Acrylonitrile.
- Propylene oxide.
- Cumene.
- Acrylic acid.
- Polypropylene.

Polypropylene is responsible for a major part of the propylene demand, followed by acrylonitrile and propylene oxide.

Propylene is normally produced in three commercial grades: refinery grade, with purity varying from 50% to 70%; the refinery grade is applied in the production of cumene, for example. The chemical grade (92% to 96%) was applied to produce acrylonitrile, propylene oxide, and acrylic acid, and finally, the polymer grade (up to 99.5%) was applied to produce polypropylene.

The main sources of propylene are the steam cracking process, the FCC units in crude oil refineries, olefins metathesis, propane dehydrogenation, and methanol to olefins processes.

The steam cracking units are the main propylene supplier, followed by the fluid catalytic cracking (FCC) units in crude oil refineries.

According to some recent forecasts, the petrochemical market tends to rise in the next few years and, in the middle term, will be responsible for a major part of the crude oil consumption, surpassing the transportation fuels. This fact has made the refiners look for closer integration with petrochemical assets through the maximization of petrochemical intermediates in their refining hardware as a strategy to ensure better refining margins and higher value addition to the crude oil.

Normally, the LPG produced in FCC units contains close to 30% of propylene, and the added value of the propylene is close to 2.5 times that of the LPG. According to the local market, the installation of propylene separation units presents an attractive return on investment. Despite the advantage, a side effect of the propylene separation from LPG is that the fuel stays heavier, leading to specifications issues, mainly in colder regions. In these cases, alternatives are to segregate the butanes and send this stream to the gasoline pool, add propane to the LPG, or add LPG from natural gas. It's important to take into account that some of these alternatives reduce the LPG offer, which can be a severe restriction according to the market demand.

A great challenge in the propylene production process is the propane and propylene separation step. The separation is generally hard by simple distillation because the relative volatility between propylene and propane is close to 1.1. This fact generally leads to distillation columns with a high number of equilibrium stages and high internal reflux flow rates.

There are two technologies normally employed in propylene-propane separation towers that are known as heat pump and high-pressure configurations.

The high pressure technology applies a traditional separation process that uses a condenser with cooling water to promote the condensation of top products. In this case, it's necessary to apply sufficient pressure to promote the condensation of products in the ambient temperature. Furthermore, the reboiler uses steam or another available hot source. The adoption of high-pressure separation route requires a great availability of low pressure steam in the refining hardware. In some cases, this can be a restrictive characteristic, and the heat pump configuration is more attractive, despite the higher capital requirements.

The separation process applying heat pump technology uses the heat supplied by the condensation of top products in the reboiler. In this case, reboiler, and condenser are the same equipment. To compensate for the non-idealities, it's necessary to install an auxiliary condenser with cooling water.

The application of heat pump technology allows a decrease in the operating pressure by close to 20 bar to 10 bar; this fact increases the relative volatility of propylene-propane, making the separation process easier and, consequently, reducing the number of equilibrium stages and internal reflux flow rate required for the separation.

Normally, when the separation process by distillation is hard (with relative volatilities lower than 1.5), the uses of heat pump technology are more attractive.

Furthermore, some variables need to be considered during the choice of the best technology for the propylene separation process, such as the availability of utilities, temperature gap in the column, and installation cost.

Normally, the propylene is produced in the refineries with specifications. The polymer grade is the most common and has higher added value with a purity of 99.5% (minimum); this grade is directed to the polypropylene market. The chemical grade where the purity varies between 90% and 95% is normally directed to other uses.

Propylene is another common product stored in the spheres tank farm of the refinery, which demands special attention for handling during the routine operations of the stockpiling area, like transfer and expedition from the spheres to the final consumers.

11.3 THE HAZARDS TO STORAGE LIQUEFIED PETROLEUM GASES

Due to their characteristics, liquefied gases like LPG, propylene, and butane tend to concentrate near the floor during leakages, causing the formation of highly flammable vapor clouds. Another very important characteristic of the liquified gases is that these compounds tend to produce great temperature reduction during expansion, leading to the risk of freezing facilities like valves and samplers.

The large inventory of LPG storage vessels is another significant factor to be considered in the risk analysis of LPG storage farms. This characteristic of the LPG is one of the main reasons that the LPG storage farms should rely on strict access control of motorized vehicles once a simple leak can produce a vapor cloud that can be inflamed under the presence of a combustion engine.

For this reason, the stockpiling operating personnel needs to be trained to develop total intolerability to LPG leaks in the operating systems of LPG storage facilities, once an LPG leakage occurs. The operating system should be immediately removed from operation, and the risk-mitigating actions should be taken until adequate maintenance service is planned and executed.

Another intrinsic risk of the LPG storage zone is the occurrence of BLEVE phenomena. BLEVE is the acronym for boiling liquid expanding vapor explosion, which makes reference to a kind of explosion that occurs when a vessel containing a pressurized liquid blows up during a fire.

The relieved gases are superheated and produce a large fireball which can enlarge the neighboring region. This phenomenon causes a drastically worsening of an emergency scenario in the LPG storage zone, which will raise the risk of the blast wave due to the explosion, the projection of fragments from the LPG sphere, and the great thermal radiation from the fireball, which can reach the neighboring spheres, creating a kind of domino effect which can reach the whole LPG spheres farm.

Unfortunately, we have had great process safety accidents occur in LPG storage farms in the history of the refining industry, which illustrates the severity of process safety accidents in LPG storage facilities and should be constantly discussed with the stockpiling operation team to call attention to how critical a leakage in these assets can be.

One of the most studied and known process safety accidents involving LPG storage farms occurred on 4th January 1966 in the Feyzin refinery at Lyon, France.

The accident occurred during an operation to collect an LPG sample from the bottom of one of the LPG spheres. During the draining process of the LPG sphere, an explosion occurred, and the LPG blew out. Under this scenario, the operator was not able to close the draining valves sufficiently quickly. The valve was frozen due to the LPG expansion, and the LPG leakage became out of control. The draining system of the LPG spheres of the Feyzin refinery is designed to drain the sphere's bottom to a sump connected to the wastewater treating system of the refinery. The draining and purge system of the Feyzin refinery is illustrated in Figure 11.1.

The operators sounded the emergency alarm, taking care to block the vehicle's transit around the leakage. The emergency response team tried to stop the LPG leakage for around ten minutes without success. Once the two draining valves were frozen, according to the accident report, an LPG cloud around 1.5 meters of thickness was formed, flowing in the direction of the motorway. The emergency response team tried to stop the traffic on the motorway. Unfortunately, a car entered the LPG cloud despite the signals from the emergency team to stop; this car stopped 160 m from the leakage, and the cloud was ignited by a hot point in the car.

The flame moves quickly in the direction of the leakage, producing a kind of torch, which is supplied by the inventory of the LPG sphere. The emergency structure is activated, and the emergency team of the refinery is reinforced with external firefighters, despite the constant efforts to contain the flames with water and foam. The process safety valve (PSV) of the LPG sphere under the leak is open, provoking

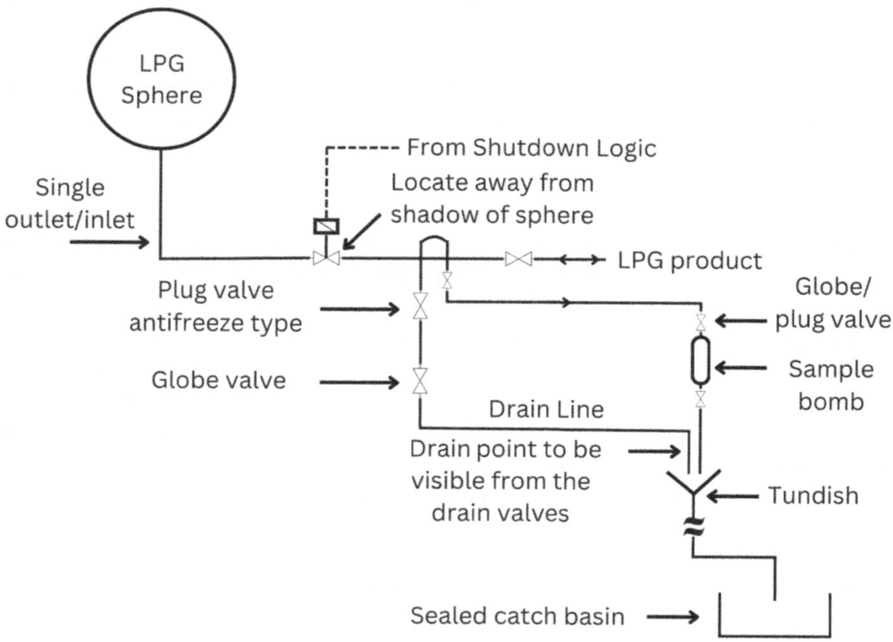

FIGURE 11.1 Schematic representation of the draining and purge device of the LPG spheres at Feyzin refinery (At the moment of the accident).

a 10-meter torch. The emergency team passes to concentrate the efforts on cooling the neighboring spheres.

Despite the opening of the PSV and the intense firefighting, the LPG sphere explodes, causing a large fireball (BLEVE phenomenon), and the emergency shutdown of the processing units is ordered. In a 10-minute interval, the neighboring spheres explode, relieving even more flammable gases to the fire. Despite the efforts of the emergency team, the fire reaches and destroys the entire LPG storage farm. Unfortunately, 18 people lost their lives in this accident.

The Feyzin accident illustrates perfectly the attention that needs to be devoted to the LPG storage zone of a crude oil refinery and calls attention to design criteria of draining and purging systems. A key factor which can be considered as the cause of the accident is the draining and purge system of the LPG spheres, which demands the operators to relieve liquefied gases to the atmosphere during the sampling process, which is a routine task. As previously mentioned, the release of LPG to the atmosphere is an unthinkable practice nowadays.

A good practice to reduce the risks during the draining and sampling processes of liquefied gas spheres is to design this system to drain and relieve the accumulated liquid and eventual gases to the flaring system, as illustrated in Figure 11.2.

Some do not agree to send deliberate liquid to the flaring system, but in this case, the amount of liquid tends to be relatively low and can be adequately separated from the flaring gases in the blow down drum, as discussed in Chapter 6.

Here, it's important to highlight that the process of draining LPG spheres to the flare system should be adequately controlled and monitored continuously by an operator, aiming to avoid the drastic temperature reduction in the blow down vessel. A temperature

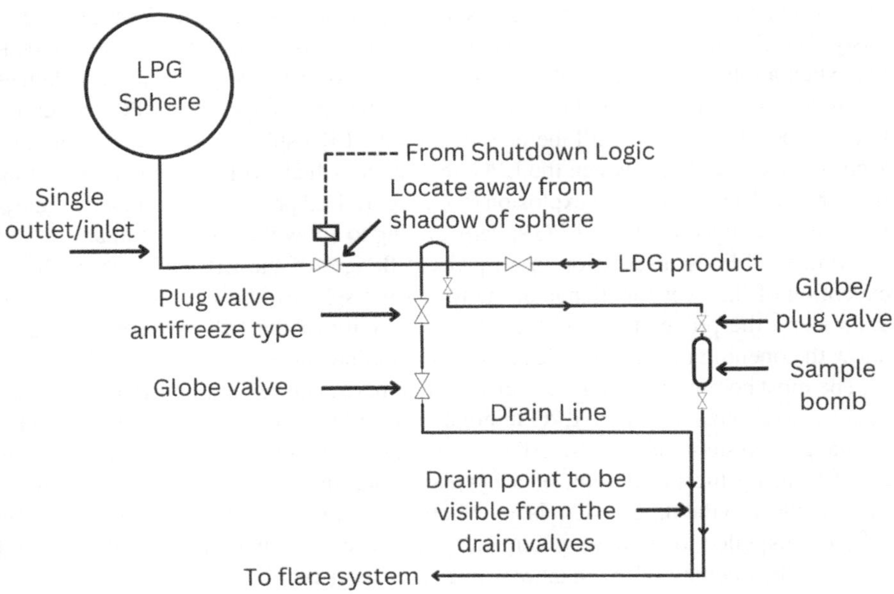

FIGURE 11.2 Draining and purge system of LPG spheres connected to flaring system.

reduction in this system can raise the solubility of the gases in the liquid phase inside the vessel, leading to the efficiency loss and cavitation of the blow down pumps, leading to a great risk of sending liquid to the flare stack. To avoid this scenario, the LPG spheres draining process to the flare system should be carried out under low flow rates, measuring the temperature in the liquid phase of blow down vessel, and under constant monitoring as previously mentioned.

The scenario mentioned above can occur during the shutdown of the LPG spheres to plan maintenance and internal inspection.

11.4 MAINTENANCE OF LPG SPHERES

Like any process equipment and storage tank, the LPG spheres demand shutdowns in determined intervals (each 5–6 years) to proceed with a detailed internal inspection to ensure the integrity of equipment for a new operating cycle.

The shutdown of LPG spheres is one of the most critical tasks managed by the stockpiling operation team and demands adequate attention and procedures to minimize the process safety risks.

The LPG sphere shutdown involves the necessity to isolate the sphere from the other spheres in the storage zone, aiming to avoid contamination and process safety risks to the maintenance personnel. This isolation is reached through the insertion of line blinds, aiming to isolate the sphere from the process and other spheres, which will be kept under operation.

The operation to insert the line blinds in LPG or other liquefied gases, such as propylene systems, should be conducted carefully and under adequate procedures. The stockpiling team should be trained to never allow the opening of LPG tubulation flanges without the certainty that there is no LPG in the tubulation, as described previously. The characteristics of LPG and other liquefied gases make it very hard to identify when a tubulation is free of LPG and there is no explosivity, making the flanges opening a safe task. To avoid the risk of opening the pipe flanges with LPG concentration, the best practice is to fill the pipe or even the LPG sphere with water, creating a kind of water blanket between the LPG and flange, which will be opened, protecting the personnel against fire and explosion risks. Figure 11.3 presents a schematic arrangement of water injection to an LPG sphere, aiming to allow the blinds installation.

After the blinds installation in all process flanges of the LPG spheres (with the exception of the overflow flange at the top of the sphere), the stockpiling operation team starts the procedure to reduce the explosivity of the LPG sphere, aiming to allow the opening of the manholes to start the maintenance services properly.

The most common procedure to ensure that there is no remaining LPG and consequently no explosivity in the LPG internals is to fill the LPG sphere with water until the gauging system indicates overflow to the flare system; normally, this overflow is carried out by the installation of a bypass device in the PSV of the LPG sphere. The overflow to the flare system is an effective way to ensure that the LPG inside the sphere is expelled to the flare system, keeping the sphere internally free of LPG, and allowing the opening of the manholes in a safer way(Figure 11.4).

After the manhole opening, the internal inspection and maintenance started with a detailed inspection of welded connections, flanges, and sphere legs, which normally

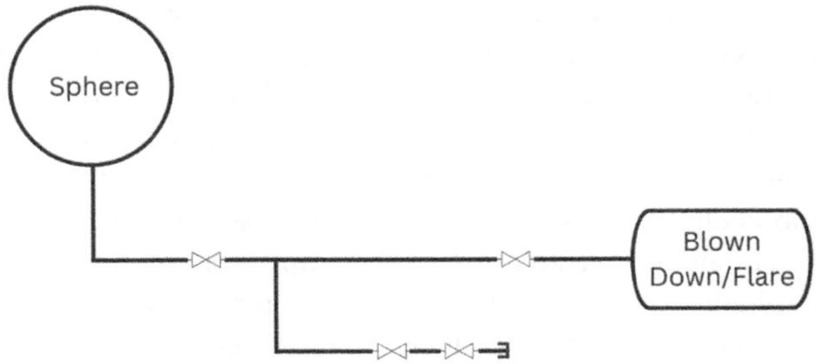

FIGURE 11.3 Schematic arrangement for blind installation in flanges of LPG spheres.

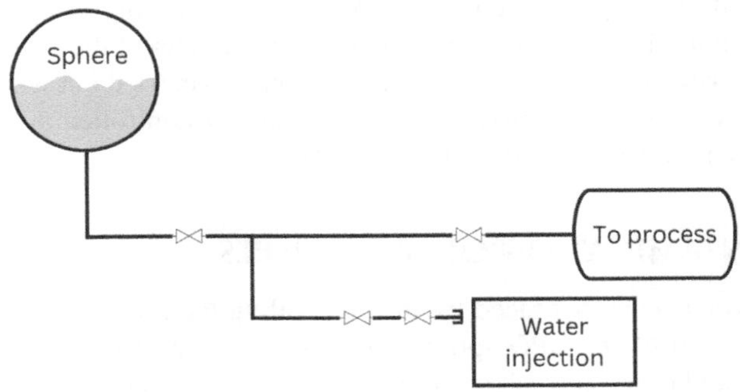

FIGURE 11.4 Schematic arrangement for the inertization process of LPG spheres.

involves a deep analysis of the fireproofing conditions. Another key question for the LPG spheres is the internal coating, which creates a barrier between the hydrocarbons and the sphere shell. The condition of the internal coating should be adequately analyzed once this coating or paint is responsible for controlling the corrosion rate of the sphere shell, which has a great impact on the lifecycle of the sphere. Normally, an LPG sphere is kept under operation for around 5 years and then is put out of service for internal inspection. The stockpiling team should be able to identify any operating restrictions and reliability threats in the operating system of the LPG spheres and demands from the maintenance team that these failures need to be corrected during the internal inspection and maintenance window of the equipment. This is especially true for the motorized valves which compose the process alignments of the LPG sphere; it's unthinkable to operate a recently inspected sphere operating process valve manually.

After the conclusion of maintenance and inspection services, the return of the LPG sphere to operation demands great attention from the stockpiling operation team. Back to operating an inspected sphere demands the removal of line blinds, and the procedure should be the same as the installation, injecting water in the pipes or in the sphere to allow a safe operation to remove the blinds of the pipes connected with the process.

Before starting the operation of the LPG sphere, or allowing the LPG supply to the sphere, it's necessary to displace the oxygen inside the sphere, aiming to avoid the formation of an explosive mixture inside the sphere which can cause an explosion. Normally, oxygen displacement is achieved by injecting inert gas (nitrogen) in the sphere, aiming to ensure that the oxygen is displaced to the flare system; the nitrogen injection is kept until the samples in the bottom and top of the sphere indicate oxygen concentration lower than 1.0%.

After this procedure, the sphere is delivered to operation, and the blasting procedure is started, equalizing the pressure of the empty sphere with the vapor phase of the operating spheres in order to avoid a quick vaporization of the LPG inside the sphere, which is backing from maintenance, which can cause process safety risks.

For spheres dedicated to the storage of propylene or other petrochemical intermediates (ethylene, butylene, etc.), an additional concern should be managed by the stockpiling and maintenance teams. Once the sphere is filled with water during the blinds removal, it's necessary to dry the sphere before starting its operation, considering that the humidity concentration is a quality parameter of the propylene. For this reason, the inertization process of propylene spheres is carried out with heated nitrogen, and the water concentration is another parameter controlled at the bottom and top of the sphere, further the oxygen concentration.

11.5 CONTAMINATION OF LPG SPHERES

One of the worst scenarios faced by the stockpiling team of a crude oil refinery is the contamination of the LPG spheres. The main cause of contamination of LPG spheres is related to the corrosivity of the LPG caused by operating instability of the LPG treating section of the production processing units, which are normally the fluid catalytic processing units.

Fluid catalytic cracking (FCC) is one of the main processes which give higher operational flexibility and profitability to refiners. The catalytic cracking process has been widely studied over the last few decades and has become the principal and most employed process dedicated to converting heavy oil fractions into higher economic value streams.

The installation of catalytic cracking units allows the refiners to process heavier crude oils and consequently cheaper, raising the refining margin, mainly in higher crude oil price scenarios or in geopolitical crises that can make access to light oils difficult. The typical catalytic cracking unit feed stream is gas oils from the vacuum distillation process. However, some variations are found in some refineries, like sending heavy coke naphtha, coke gas oils, and deasphalted oils from deasphalting units to processing in the FCC unit.

The catalyst normally employed in fluid catalytic cracking units is a solid constituted by small particles of alumina (Al_2O_3) and silica (SiO_2) (zeolite). To the catalyst characteristics and the operational conditions in the catalytic cracking process (temperature higher than 500°C), the process is inefficient for cracking aromatic compounds; therefore, the more paraffinic is the feed stream, the higher the unit conversion. Figure 11.5 presents a process scheme for a typical fluid catalytic cracking unit (FCCU).

In a conventional scheme, the catalyst regeneration process consists of the carbon partial burning deposited over the catalyst, according to the chemical reaction below:

$$C + ½O_2 \rightarrow CO$$

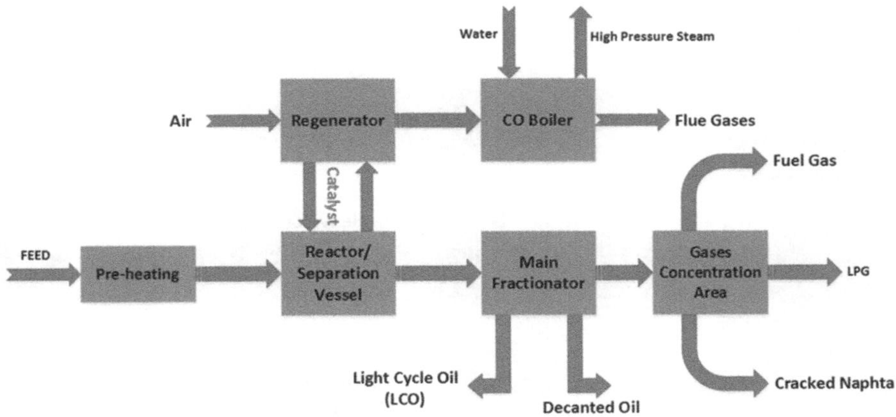

FIGURE 11.5 Schematic process flow for a typical fluid catalytic cracking process unit (FCCU).

The carbon monoxide is burned in a boiler capable of generating higher pressure steam that supplies other process units in the refinery.

The principal operational variables in a fluid catalytic cracking unit are reaction temperature, normally considered the temperature at the top of the reactor (called a riser), feed stream temperature, feed stream quality (mainly carbon residue), feed stream flow rate, and catalyst quality. Feedstock quality is especially relevant, but this variable is a function of the crude oil processed by the refinery, so it is difficult to change, but, for example, aromatic feedstock with high metals content are refractory to cracking and conducting to a quick catalyst deactivation.

An important variation of the fluid catalytic cracking technology is the residue fluid catalytic cracking unit (RFCC). In this case, the feedstock to the process is basically residue from the atmospheric distillation column, due to the high carbon residue and contaminants (metals, sulfur, nitrogen, etc.) are necessary for some adaptations in the unit, like a catalyst with higher resistance to metals and nitrogen and catalyst coolers. Furthermore, it's necessary to apply materials with the most noble metallurgy due to the higher temperatures reached in the catalyst regeneration step (due to the higher coke quantity deposited on the catalyst), which raises significantly the capital investment to the unit installation. Nitrogen is a strong contaminant to the FCC catalyst because it neutralizes the acid sites of the catalyst, which are responsible for the cracking reactions.

When the residue has high contaminant content, it is common for the feed stream treatment in hydrotreating units to reduce the metals and heteroatom concentration to protect the FCC catalyst.

Typically, the average yield in fluid catalytic cracking units is 55% in volume in cracked naphtha and 30% in LPG. Figure 11.6 presents a scheme for the main fractionator of the FCC unit with the principal product streams.

The decanted oil stream contains heavier products and has high aromatic content. It is common that this product is contaminated with catalyst fines, and normally this stream is directed to use like fuel oil diluent, but in some refineries, this stream can be used to produce black carbon.

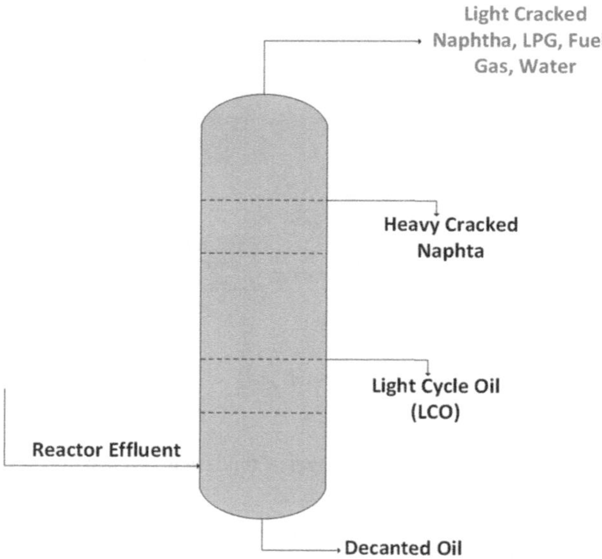

FIGURE 11.6 Main fractionator scheme for a typical fluid catalytic cracking unit.

Light cycle oil (LCO) has a distillation range close to diesel, and normally this stream is directed to treatment in severe hydrotreating units (due to the high aromaticity); after this treatment, the LCO is sent to the refinery diesel pool.

Heavy cracked naphtha is normally directed to the refinery gasoline pool; however, in scenarios where the objective is to raise the production of middle distillates, this stream can be sent to hydrotreating units for further diesel production.

The overhead products from the main fractionator are still in the gaseous phase and are sent to the gas separation section. The fuel gas is sent to the refinery fuel gas ring, after treatment to remove H_2S, where it will be burned in fired heaters while the LPG is directed to treatment (MEROX) and further commercialization.

Cracked naphtha is usually sent to the refinery gasoline pool, which is formed by naphtha produced by other process units such as straight-run naphtha and naphtha from the catalytic reforming unit. Due to the production process (deep conversion of residues), the cracked naphtha has high sulfur content, and to attend to the current environmental legislation, this stream needs to be processed to reduce the contaminant content, mainly sulfur.

Usually, catalytic cracking units are optimized to aim the production of fuels (mainly gasoline); however, some process units are optimized to maximize the light olefins production (propylene and ethylene). Process units dedicated to this purpose have his project and operational conditions significantly changed once the process severity is strongly raised in this case.

The reaction temperature reaches 600°C and a higher catalyst circulation rate raises the gas production, which requires a scaling up of the gas separation section. Figure 11.7 presents a typical scheme for a gas separation section of a fluid catalytic cracking unit.

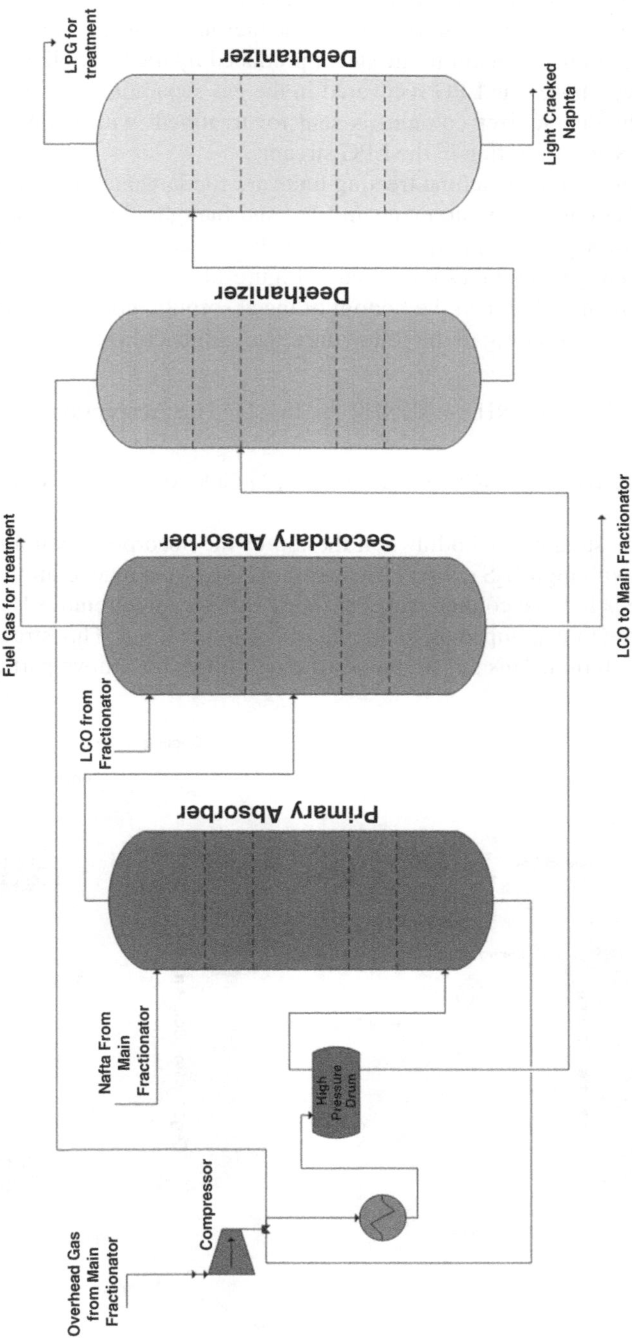

FIGURE 11.7 Basic process flow diagram for a typical gas separation section from FCC unit.

In several cases, due to the higher heat necessity of the unit, it is advantageous to operate the regenerator with the total combustion of the coke deposited on the catalyst. This arrangement significantly changes the thermal balance of the refinery once it's no longer possible to resort to the steam produced by the CO boiler.

As described above, the LPG recovered in the gas separation section of the FCC unit (top of the debutanizer column) is sent for treatment with amines, aiming to control the H_2S concentration in the LPG stream.

In the refining industry, amine treating units are fundamental to meeting the corrosivity and H_2S content requirements in LPG and fuel gas. The process consists of the treatment of a gas stream with an amine solution, contaminant absorption, and amine solution regeneration, as presented in Figure 11.8.

The feed stream is fed in at the bottom of the absorption column, where H_2S and CO_2 are removed according to the following chemical reactions:

$$H_2S + R2NH \leftrightarrow R2NH2^+ + HS^- \quad (1)\ H_2S\,\text{Absorption}$$

$$CO_2 + 2R2NH \leftrightarrow R2NCOO^- + R2NH2^+ \quad (2)\ CO_2\,\text{Absorption (Simplified)}$$

The treated gas stream is withdrawn at the top of the absorption column while the amine stream with high H_2S content (rich amine) is sent to a regeneration step where the rich amine is fed to a column and is stripped with steam to remove H_2S and CO_2. The bottom stream is pumped again to the absorption column. This stream is called a lean amine solution. This stream passes through filters to remove particulates and

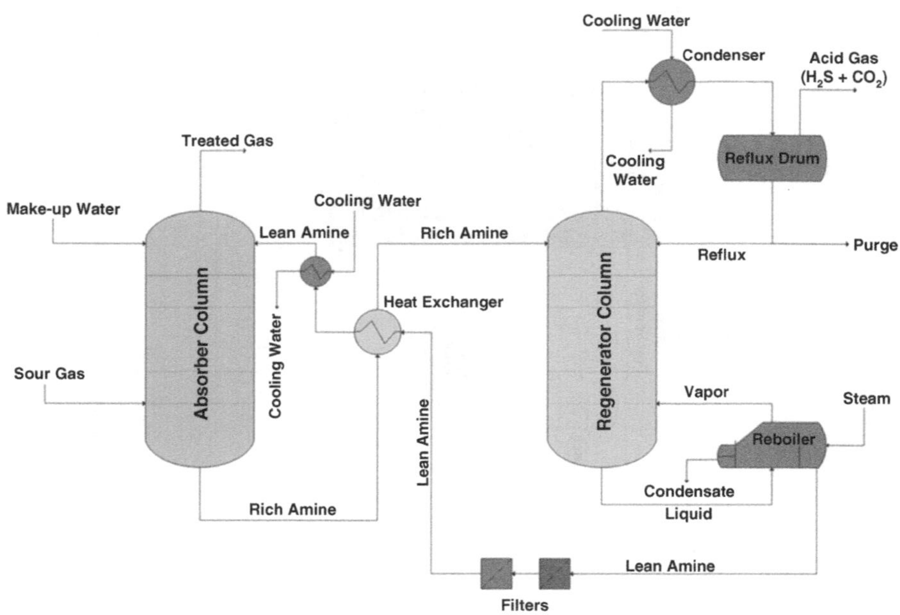

FIGURE 11.8 Basic process scheme for a typical amine treating process unit.

impurities before reaching the absorption tower. The lean amine exchanges heat with the rich amine stream.

The stream recovered at the top of the regeneration tower is called sour gas stream and has a high H_2S content. Normally, this stream is directed to sulfur recovery units where the H_2S is converted into elemental sulfur through the Claus process. For units dedicated to treating LPG, there is a feed stream washing step (with water) to remove insoluble compounds (mercaptans, thiocyanide, etc.).

The main process variables of the amine treating process are pressure in the regeneration column (1.5 to 1.8 bar), the temperature of amine solution in the absorption and the regeneration columns, temperature in the top of regeneration column, H_2S content in the lean amine stream, and the temperature in the bottom of the regeneration column (115 to 125°C).

An important control parameter in the process is the solids content in the lean amine stream. A high solids content can lead to a foaming formation that makes the process control difficult and leads to the drag of amine solution by the gaseous streams. To avoid solids accumulation, the lean amine stream is filtered through carbon filters (to remove organic contaminants) and cartridge filters (to remove inorganic solids). Other relevant variables are the amine circulation flow rate and free amine content. In other words, the quantity of amine that doesn't suffer thermal or chemical degradation during the process. Normally, the fixed mine content (that suffered degradation) is controlled below 5%.

The amine type that is applied also has a profound influence on the process results; amines normally employed are MEA (Monoethanolamine), DEA (Diethanolamine), and MDEA (Methyl-Diethanolamine). MEA is normally applied in natural gas treatment once the acid compounds content is lower. Due to high chemical activity, the MEA is normally harder to recover when compared with DEA, which is commonly used in refinery treating units. The MEA is applied in solutions with 15 to 20% in mass, while DEA is applied in solutions with 20 to 30% in mass. MDEA presents a lower H_2S-removing efficiency and normally is not applied to remove CO_2. On the other hand, due to the lower activity, it is easily regenerated when compared with MEA and DEA; however, MDEA has a higher cost, and a greater solubility in hydrocarbons leads to losses and a high makeup flow rate. MDEA is normally employed in solutions with 40–50% in mass.

Regenerative caustic treatment applies the continuous regeneration of caustic solution. This significantly reduces soda consumption in the process, leading to a reduction in operational costs and making the caustic process more flexible in relation to feed stream contaminant concentration. Regenerative caustic treatment can be applied to LPG, naphtha, and kerosene (jet fuel).

The chemical reactions involved in the regenerative caustic treating process are presented as follows:

$$RSH + NaOH = NaSR + H_2O \tag{11.1}$$

$$2\,NaSR + H_2O + \tfrac{1}{2}O_2 \rightarrow RSSR + 2\,NaOH \tag{11.2}$$

Reaction (11.2) occurs in the presence of a homogeneous cobalt-based catalyst and promotes the caustic solution regeneration applied in the mercaptans extraction.

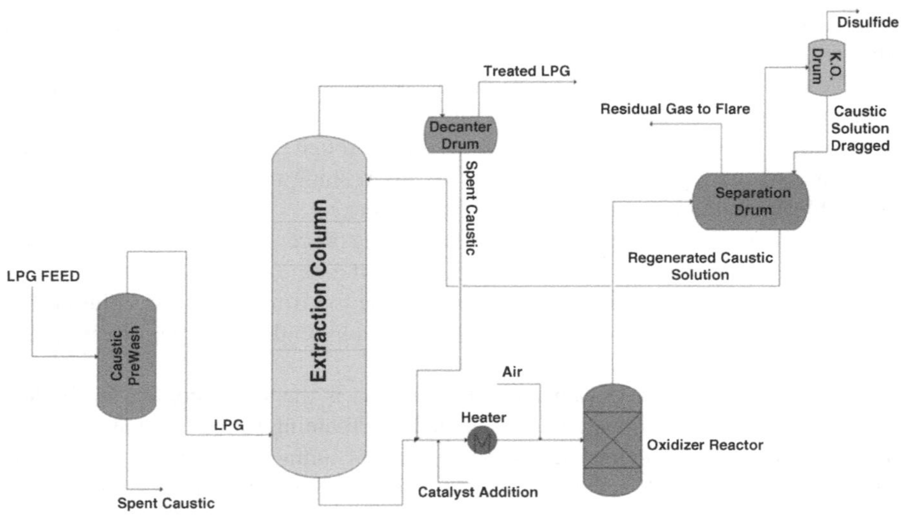

FIGURE 11.9 Typical process arrangement to LPG regenerative caustic unit.

Figure 11.9 presents a process flow diagram of a typical regenerative caustic process unit dedicated to treating LPG.

LPG is fed to a caustic pre-wash drum, aiming to remove H_2S and contaminants that can prejudice the treatment process. In the next step, the LPG is directed to an extraction column where the contact between LPG and caustic solution occurs. Treated LPG is recovered in the top of the column, and the spent caustic solution is pumped to the treatment system from the column bottom.

The bottom stream containing the spent caustic solution is heated and receives an air injection before being fed to an oxidizer reactor where the extracted mercaptans are oxidized. After this step, the stream is sent to a separation drum, from where the regenerated solution is pumped back to the extraction column and the disulfides are sent to the refinery fuel gas pool. Nowadays, due to the need to attend to the environmental regulations related to atmospheric emissions, this stream is usually sent to hydrotreating units.

LPG regenerative caustic process promotes the sulfur compounds extraction once the mercaptans are removed in the aqueous phase. In the case of naphtha and Jet fuel treatments, the mercaptans are converted to disulfides but kept in the hydrocarbon. In these cases, the regenerative caustic process is called the sweetening process; the total sulfur content in the product does not change, but the derivative becomes less corrosive.

The main process variables of the LPG regenerative caustic process are the temperatures of the extraction column and oxidizer reactor, air and caustic solution flow rates, caustic solution density, and the operating pressure of the extraction column that is controlled in the range to keep the LPG in the liquid phase (13–14 kgf/cm^2).

An operational instability of the LPG treatment section can raise the sulfur content of the LPG in case of poor performance of the caustic regenerative step and the corrosivity issues in case of poor performance or operating issues of the LPG amine treating.

The total sulfur issue is an easier issue to deal with in the LPG storage farm once the sulfur content can be optimized through dilution and the LPG inventory of a sphere with high sulfur content can be diluted with low sulfur production, just opening enough space to proceed with the correction of the total sulfur concentration in the final inventory.

The corrosivity is the worst scenario and is hard to deal with in the stockpiling operations once the corrosivity can't be corrected through dilution, and the H_2S, responsible for the corrosivity, tends to concentrate in the vapor phase of the LPG spheres. If the contaminated sphere is not adequately isolated, the whole LPG storage farm can be contaminated.

The copper corrosion test is one of the various analytical tests applied in the LPG analysis to ensure the quality of this derivative. The objective of the copper corrosion test is to identify the presence of acid gases, mainly H_2S, which can be deleterious to the burners. The test consists of putting the LPG in contact with a copper coupon under standard conditions and checking the color of the copper coupon with standards; in the LPG case, the copper corrosion test is made following the ASTM D1838-16 standard.

Elemental sulfur and reactive sulfur species like hydrogen sulfide and mercaptans will react with the surface of the copper coupon, leading the LPG sample to fail the test (corrosion higher than 1 according to the standard). The carryover of caustic solution or amine can be another reason for LPG copper corrosion failure. This is a relatively common situation during the startup of fluid catalytic cracking units, where the unstable operating conditions can lead to the poor performance of the amine treating and caustic regenerative treating steps of the LPG and cause the sending of corrosive LPG to be sent to the spheres storage farm.

As previously mentioned, once an LPG sphere contaminated with corrosive LPG is identified, this sphere should be isolated from the others through the insertion of blinds in the process pipes, and the pressure of the contaminated LPG sphere should be kept lower than that of the other spheres, aiming to avoid the contamination of the other LPG spheres through the vapor phase. This can be achieved by aligning the vapor phase of the contaminated sphere to the flaring system.

Before isolating the contaminated LPG sphere, it's important to ensure that the causes of the LPG contamination in the production unit are identified and eliminated, aiming to avoid the alignment of corrosive products for a clean sphere. Once a sphere is already contaminated, the LPG production should be kept to this sphere until two consecutive samples of the LPG production indicate that the production is under control.

Once the contaminated sphere is isolated, there are two ways to ensure the decontamination of the equipment:

- Opening the equipment and cleaning the shell with pressurized water: This is a good alternative if the LPG sphere is close to their internal inspection or if the inspection is anticipated. Otherwise, this strategy can be difficult to execute, as it demands a lot of maintenance resources and exposes the employees to significant process and occupational risks.

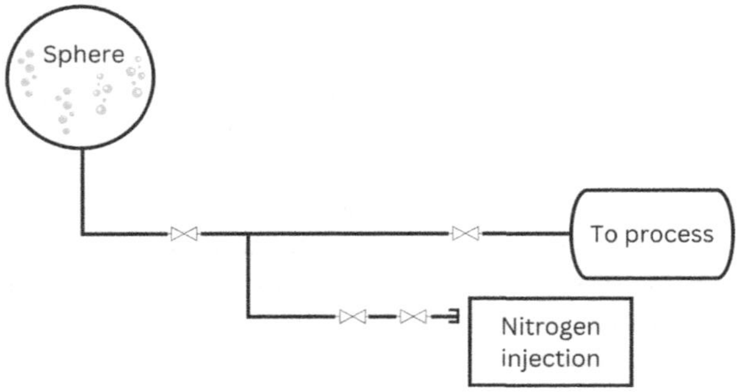

FIGURE 11.10 Schematic arrangements for the decontamination of LPG sphere using water injection.

- Washing the internal shell of the sphere with water injection: This strategy can be applied with adequate procedures through the water injection to the contaminated sphere, followed by the water draining, as presented in Figure 11.10.

To ensure that the process will be effective, the contaminated sphere should be filled with water and drained twice. After this procedure, the LPG sphere can be placed under operation again to storage clean LPG without issues of corrosive LPG.

11.6 PROPYLENE CONTAMINATION WITH CARBONYL SULFIDE (COS)

Another contamination case of the liquefied gases storage farm commonly managed by the stockpiling operation personnel is propylene contamination with carbonyl sulfide (COS).

The carbonyl sulfide is a poison to the Ziegler Natta catalysts applied in the petrochemical assets which apply propylene as an intermediate. Normally, the maximum concentration of carbonyl sulfide in the polymer grade propylene is less than 20 ppbv (parts per billion by volume). This is the first difficulty faced by the refiners in controlling the COS content in the propylene once the refining process is not designed to deal with the control of very low contaminant concentrations as practiced in the COS control.

Carbonyl sulfide production occurs during the catalytic cracking reactions in the FCC units, and some references quote that a major part of the COS is produced in the catalyst regeneration section and is dragged to the FCC riser in the pores of the catalyst. Once the carbonyl sulfide boils slightly below the propane, the COS tends to concentrate in the mixture of propane/propylene, which is fed to a propylene recovering unit.

The COS formation is strictly related to the sulfur content in the feed of the FCC units as well as the process severity, like reaction temperature (TRX) and the catalyst/oil relation. The greater the operational severity of the FCC unit, the higher the concentration of the carbonyl sulfide in the propane/propylene mixture; this is especially true for partial combustion FCC units.

There is no consensus about the exact production mechanism of COS in the FCC units, but it is expected that the mechanism involves the chemical reactions below.

$$H_2S + CO_2 \leftrightarrow COS + H_2O$$

$$2\,CO + 2\,S \leftrightarrow 2\,COS$$

$$SO_2 + 3\,CO \leftrightarrow COS + 2\,CO_2$$

As previously mentioned, the COS concentration depends on the sulfur content in the FCC feed and the operational severity of the unit. For high sulfur feeds, the use of catalyst additives like ZSM-5 is preferred against higher severity to maximize the propylene yield in order to minimize the problems related to high COS concentration.

Despite the contribution of amine and caustic treating to reduce the COS concentration, these processes are not effective in controlling the carbonyl sulfide concentration in the propylene. Normally, the COS concentration is controlled through hydrolysis of the COS to H_2S or adsorption methods.

A typical FCC propylene separation processing unit is presented in Figure 11.11 below.

The LPG (C3 mixture) from the FCC unit is pumped to a depropanizer column where the light fraction (essentially a mixture of propane and propylene) is recovered in the top of the column and sent to a deethanizer column while the bottom (butanes) is pumped to LPG, or gasoline pool, according to the refining configuration. The top stream of the deethanizer column (the lighter fraction) is sent back to the FCC, where it is incorporated into the refinery fuel gas pool, or in some cases, can be directed to petrochemical plants to recover the light olefins (mainly ethylene) present in the stream, while the bottom of the deethanizer column is pumped to the C3 splitter column, where the separation of propane and propylene is carried out. The propane recovered at the bottom of the C3 splitter is sent to the LPG pool, where the propylene is sent to the propylene storage park. The feed stream passes through a caustic wash treatment, aiming to remove some contaminants like carbonyl sulfide (COS). The COS is hydrolyzed as presented below.

$$COS + H_2O \leftrightarrow H_2S + CO_2$$

The caustic solution is applied to chemically trap the CO_2, H_2S, and mercaptan, as presented below:

$$H_2S + 2\,NaOH \leftrightarrow Na_2S + 2\,H_2O$$

$$RSH + NaOH \leftrightarrow RSNa + H_2O$$

$$CO_2 + 2\,NaOH \leftrightarrow Na_2CO_3 + H_2O$$

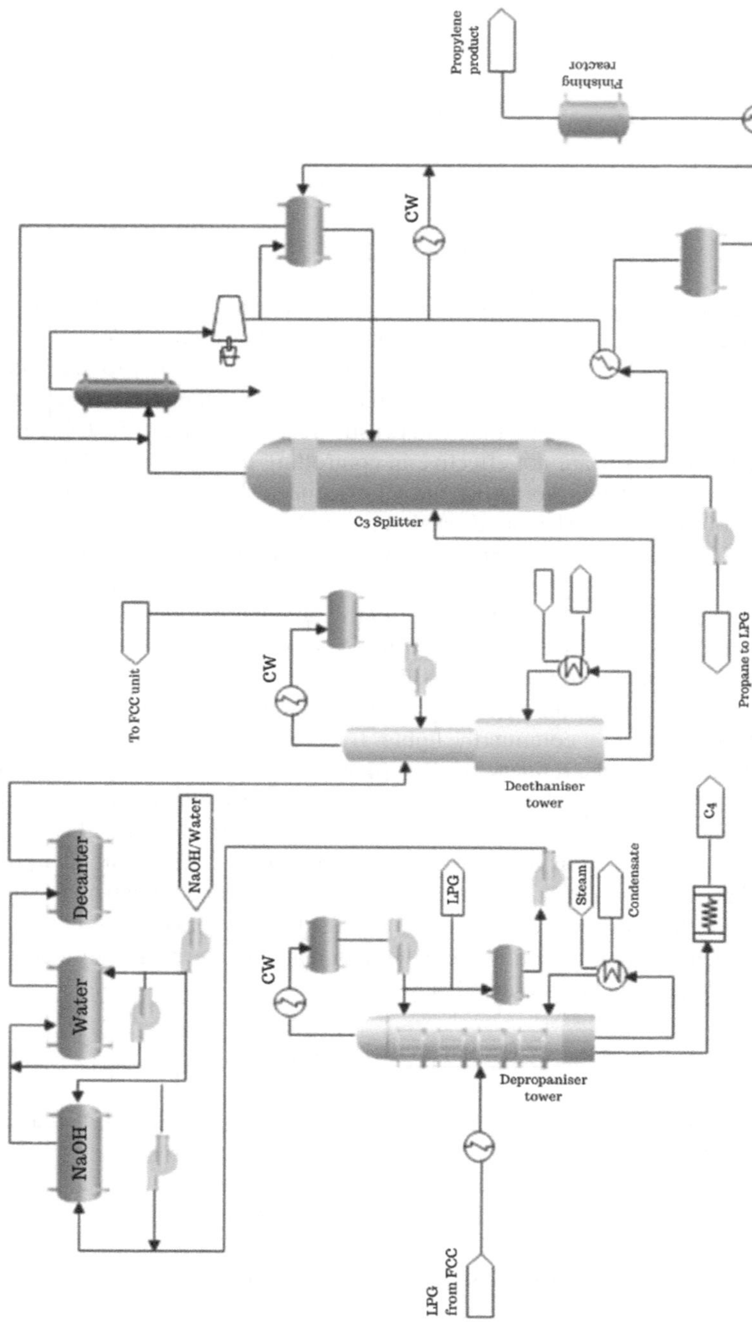

FIGURE 11.11 Typical FCC propylene separation unit.

The reaction products are water soluble and are drained with the spent caustic solution. The caustic concentration is an important process variable and should be monitored continuously by the optimization team of the refinery, aiming to ensure the performance of the treating section of the propylene recovery unit. This is also very important to avoid an overload of the finishing reactor of the unit, which is responsible for controlling the remaining COS content and arsine (AsH_3) concentration in the propylene stream, as presented in Figure 11.11.

The arsine control in the propylene is important once a high concentration of this contaminant can lead to high consumption of polymerization catalysts in the petrochemical plants. The arsine treater is normally composed of a bed containing copper oxide, which reacts with the arsine, removing the contaminant from the propylene.

Copper oxide is also effective in removing sulfur contaminants like COS, H_2S, and mercaptans (RSH), as presented below:

$$COS + CuO \rightarrow CuS + CO_2$$

$$RSH + CuO \rightarrow CuS + ROH$$

$$H_2S + CuO \rightarrow CuS + H_2O$$

The copper oxide bed has a limited capacity to adsorb the sulfur and arsine contaminants, and their saturation needs to be monitored by the process engineering team of the refinery, aiming to allow adequate planning to shut down the propylene separation unit and replace the copper oxide catalyst with minimum impact on the production planning of the refinery as well as the market supply.

As previously mentioned, the monitoring of the amine and caustic treating sections of the FCC unit, as well as the caustic washing section of the propylene recovery unit, is fundamental to avoid an overload of contaminants and consequently a short lifecycle of the copper oxide bed, which can lead to operating issues like the contamination of the propylene spheres with high concentrations of COS.

It's important to note that higher concentrations of COS in the propylene can also produce higher concentrations of CO_2, which is another quality parameter of the propylene (normally controlled below 2.0 ppm). Once as presented above, the COS hydrolysis produces CO_2. To overcome the high CO_2 concentration in the propylene spheres, it's possible to reduce the operating pressure of the deethanizer column or, in extreme cases, relieve the top vessel of the deethanizer column to the flaring system, aiming to relieve the COS and CO^2 "trapped" in this system.

11.7 WATER FLOODING SYSTEM OF LIQUEFIED GASES STORAGE FARM

One of the main safety systems of a liquefied gas storage farm is the flooding system. This system is applied in emergency scenarios, as presented in the Feyzin case study discussed previously, aiming to allow a quick water injection in the liquefied gas

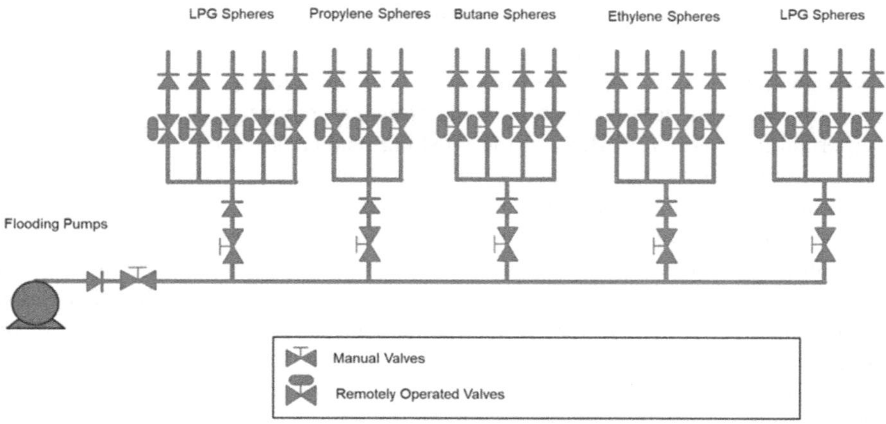

FIGURE 11.12 Schematic arrangement for a water flooding system installed in liquified gases storage farms.

spheres, minimizing the risks of fire and BLEVE occurrences. Figure 11.12 presents a typical arrangement for a water flooding system installed in a liquefied gas storage farm.

The API 2510 presents the general instructions for the adequate design and operation of water flooding systems of liquefied gas storage farms. It's important to consider the risks associated with water freezing, especially if the water injection is made at the bottom of the spheres to the detriment of the top of the spheres. Another key point, as presented in Figure 11.12, is the installation of check valves with adequate leak grades capable of ensuring that the water system will not be contaminated with liquefied gases, which can be very dangerous, mainly under an emergency scenario.

Considering the relevance of the water flooding system for the safety of the liquefied gas storage farm, it's fundamental to take action to ensure high levels of reliability and availability of the water flooding systems. Normally, the water flooding systems are designed with two pumps, where just one is capable of supplying the water flowrate injection to overcome the limiting case, which is normally defined by the lighter liquefied gas spheres (in Figure 11.12, the ethylene spheres will be the delimiting case), and the second pump is kept in the reserve condition, and, due to the reliability requirements, the pumps are actioned by different power sources; for example, one pump is driven by electric power while a second pump is driven by a diesel engine.

The water flooding operating system should rely on adequate reliability and maintenance plans, which need to be based on special materials supply plans, aiming to ensure adequate agility to the maintenance. Furthermore, the refinery personnel (HSE, maintenance, and stockpiling teams) need to carry out frequent emergency tests, aiming to identify failures and keep the operation team adequately trained to operate this system in a safe and agile manner.

BIBLIOGRAPHY

American Petroleum Institute, *Design and Construction of LPG Installations*, API Recommended Practice 2510, 9th Ed., June 2024.

Fahim, M.A.; Al-Sahhaf, T.A.; Elkilani, A.S. *Fundamentals of Petroleum Refining*. 1st ed. Elsevier Press, 2010.

Moulijn, J.A.; Makkee, M.; Van-Diepen, A.E. *Chemical Process Technology*. 2nd ed. John Wiley & Sons Ltd., 2013.

Myers, R.A. *Handbook of Petroleum Refining Processes*. 3rd ed. McGraw-Hill, 2004.

Silva, M.W. Personal Notes.

Sutton, I. *Plant Design and Operations*. 2nd ed. Gulf Professional Publishing, 2017.

12 Management of Metering and Billing Operations

Considering their impact on society and the great number of resources which are moved, the crude oil refining industry should be strictly regulated, aiming to ensure that the resources are being adequately managed and the responsibilities and custody limits have been respected.

All players involved in the crude oil refining processing chain calculate taxes, invoices, and payments based on the crude oil derivatives received or pumped out from the refinery and an adequate management of the refinery inventory and the total volume of crude oil derivatives sold. Its fundamental do allow the refinery operation in compliance with the current regulations. Refiners around the world should keep a routine of checking their metering systems, aiming to demonstrate adequate inventory management of the crude oil refinery inventories in compliance with the Sarbanes Oxley (SOX) regulation, for example.

The way to ensure accurate metering of the volume of hydrocarbons commercialized by the refinery is normally to apply metering systems, as presented in Figure 12.1.

The analyzer instruments provide data from the hydrocarbon stream, like temperature, density, viscosity, and pressure, aiming to reach an adequate and accurate metering of the volume which is being pumped out from the refinery and delivered to the customers. It's important that the refinery relies on one metering station for each derivative, like diesel and gasoline.

The most common meter applied in the metering station is a turbine meter with dual pulse outputs. To keep the meter accurate, the meters are frequently calibrated using a prover loop, as presented in Figure 12.1.

Another common meter applied in crude oil refineries is the positive displacement meter. The positive displacement meter is among the oldest devices used to measure liquids. It is considered reliable within certain limitations. It consists of a lower section with a rotor containing vanes or impellers, a central part with gears that connect the rotor with the recorder, and, in the upper part, the recorder. The positive displacement meter is best suited for measuring very viscous products such as fuel oils and for those containing paraffin or other sediments.

In the prover loop, a calibrated ball or piston is moved through the loop, and a calibrated volume is provided between two detectors (Z). When a meter is calibrated, the four-way valve presented in Figure 12.1 opens to allow the process fluid to flow behind the ball. The number of pulses from it passes from one detector, Z, to the other and is counted. After one loop, the four-way valve turns to reverse flow direction, and

DOI: 10.1201/9781003472216-12

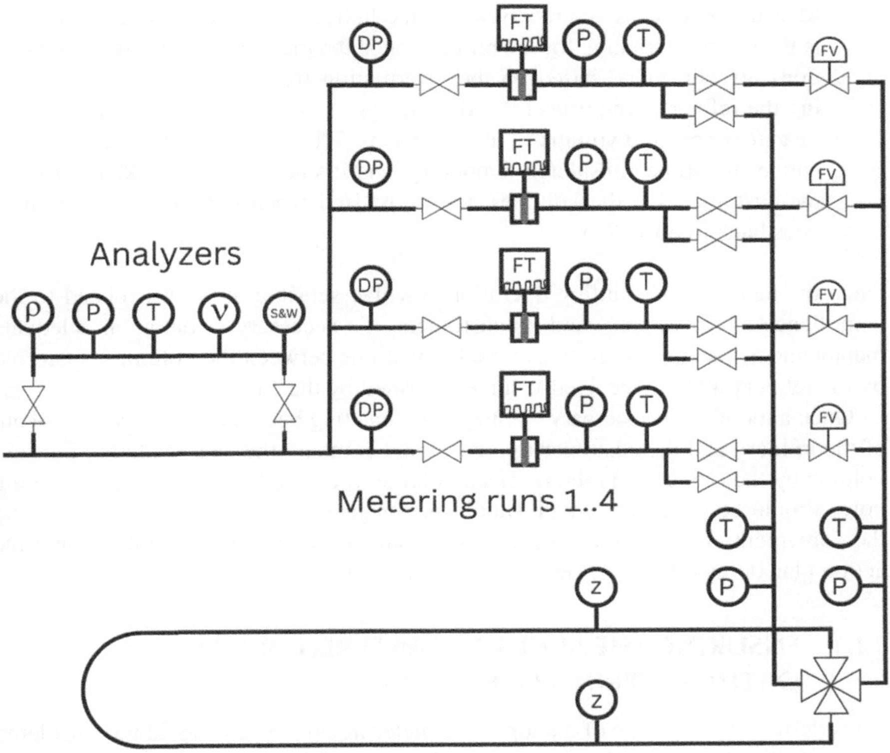

FIGURE 12.1 Example of metering system applied in crude oil refineries.

the ball moves back, providing the same volume in reverse, again counting the pulses. From the known reference volume, number of pulses, pressure, and temperature, the flow computer can calculate the meter factor and provide accurate flow measurements using formulas from industry standard organizations such as API MPMS and ISO 5024. The accuracy is typically ± 0.3% of the standard volume.

The availability of the metering stations is a key factor in the compliance of the metering requirements in stockpiling operations, and these assets demand specific maintenance plans, aiming to minimize the failures occurrences and their unavailability. When the metering stations are unavailable, there are two ways to keep a minimum compliance with the metering requirements:

- Measurement of the drop level in the sending tank – When the metering stations are unavailable, this is the preferred way to control the volume transferred to the customer. This alternative presents lower accuracy than the metering stations, which rely only on the accuracy of the gauge system of the sending tank and can lead to high discrepancies between the volumes pointed out by the refinery and the customer, creating potential conflicts. This alternative is not applicable to liquefied gases like LPG and propylene.

- Measurement using the received volume indicated by the customer – This is the worst alternative for the refinery once the metering and billing operations are controlled based on the information from the customer, exposing the refiner to inquiries related to adequate management of the refinery inventory and sold volume of hydrocarbons. When implemented, this alternative should be considered temporary, and it's necessary to make clear to the customer that the volumes will be revised when the metering station goes back to operation.

Once the metering and billing operations involve sensible questions related to the compliance of the refinery and its customers, it's necessary to define an adequate management system to control eventual deviations between the volume transferred by the refinery and the received volume informed by the customers.

Once a month, it's necessary to carry out a meeting between the stockpiling team of the refinery and the customer to present and compare the data related to the sold volume by the refinery and the received volume informed by the customer. Meeting notes should be produced, which should be assigned by the stakeholders, exposing the agreement and disagreement points related to the metering operations and the action plan defined to solve the disagreement points.

12.1 ENSURING THE ACCURACY AND RELIABILITY OF THE METERING OPERATIONS

Considering the relevance of the topic, the metering operations should be considered a priority for the stockpiling manager and it's necessary to take actions capable of ensuring the availability, reliability, and accuracy of the metering assets in order to ensure the credibility of the metering system of the refinery.

As previously mentioned, it's necessary to ensure the availability of the metering stations as well as their accuracy through periodical calibration procedures. As discussed in Chapter 8, any stockpiling asset that does not present an immediate impact on the refinery production planning shows a trend that the maintenance management system dedicates low priority to the metering assets. The stockpiling manager should be able to define an adequate management process capable of calling adequate attention to the metering system asset management.

A good strategy to achieve this goal is to develop management panels showing the availability of the metering stations, which should be presented in the management meeting of the refinery. This strategy will help the stockpiling management team to call the attention of the refinery management team to the relevance of the metering systems to the good relationship with the refinery customers as well as with the compliance with regulatory requirements like the Sarbanes-Oxley law.

The metering stations should suffer calibrations in frequent intervals, aiming to ensure the accuracy of the instruments and the reliability of the metering systems. Normally, the metering turbine is calibrated monthly while other instruments are calibrated every six months. The calibration certificates should be archived in a traceable and auditable manner in order to facilitate access to these documents during internal and external audits.

Of course, another key operational system which is pivotal to the metering and billing process of the refinery is the tank farm.

Each storage tank should have its internal volume calibrated using an outage measurement process (tonnage) in order to allow the stockpiling team to know the storage volume as a function of the tank level; this information will be applied to calibrate the gauging system.

The tonnage or outage measurement process of each tank should be conducted at the startup of the equipment and at each internal maintenance/inspection opportunity. Again, the tonnage or outage measurement certificates should be archived in adequate ways, aiming to allow easy and quick access to these documents for calibrating and auditing purposes.

Considering the requirements of regulations like the Sarbanes-Oxley Act from 2002, the refiners are obligated to keep detailed information and registers, ensuring adequate internal control of their inventories of hydrocarbons. To ensure this compliance, beyond the outage measurement of the storage tanks, the stockpiling area of the refinery needs to carry out the calibration of the gauging system of the tanks through the manual measurement of the level and temperature of each storage tank in frequent time intervals. The frequency of the measurements relies on the purpose of the storage tank, for example:

- Crude oil storage tank – Should be measured in three different levels every six months.
- Intermediate tanks (feed tanks of processing units) – The electronic gauging system should be calibrated using hand gauging tape every six months.
- Final products tanks (gasoline, diesel, jet fuel, etc.) – When the metering stations are available, the measurement interval can be every three months; when the metering stations are unavailable, the measurement interval needs to be reduced for every month.

As previously mentioned, the hand gauging system applies a specially designed measurement tape to measure the liquid level inside the tank and is applied to calibrate the electronic gauging systems of the tank farm.

Another important measurement which needs to be carried out during the tank measurement is the water level inside the tank. Some refineries do not rely on closed-loop draining systems and can have water accumulation inside the tank. The water level measurement can be achieved using a special reactant that changes color in contact with the water that will change the color of the hand gauging tape, showing the water level inside the tank.

According to the refinery dimensions, the number of tanks is very high, and the stockpiling management team should develop a measurement plan considering each tank under operation, aiming to ensure that each tank will be measured in the correct time interval.

Monthly, the commercialization area should promote meetings with the clients where the stockpiling teams will present the conditions and reliability of the metering systems, aiming to demonstrate the robustness of the management actions taken to ensure that the metering differences are under control and in compliance with the standards and rules. It's important for the stockpiling manager to consider that the commercialization team has the tendency to favor the clients in any discussion to the detriment of the refinery management. Considering this fact, the stockpiling manager needs to keep under control the metering management system in compliance with the standards and good practices, aiming to be able to technically ensure that the process is sufficiently robust to support the commercial decisions.

BIBLIOGRAPHY

American Petroleum Institute, *API Manual of Petroleum Measurement Standards*, API Manual 14.10, 2nd Ed., 2021.

API MPMS 5.3, *Manual of Petroleum Measurement Standards – Chapter 5: Metering, Section 3 – Measurement of Liquid Hydrocarbons by Turbine Meters*, American Petroleum Institute, Washington DC, 2005.

Barua, L.D. *Natural Gas Measurement and Control: A Guide for Operators & Engineers*. 1st ed. McGraw-Hill, 1991.

NORSOK I-106, *Fiscal measurement system for hydrocarbon liquid and gas*, NORSOK standard I106, 2nd ed., 2014.

Silva, M. W. Personal Notes.

Sutton, I. *Plant Design and Operations*. 2nd ed. Gulf Professional Publishing, 2017.

13 Special Topics in Crude Oil Refining

13.1 SYNERGY BETWEEN THE FCC AND HYDROCRACKING PROCESSING UNITS AS A STRATEGY TO MAXIMIZE REFINING MARGINS

Nowadays, the capacity to add value to the bottom barrel streams represents a great competitive advantage among refiners, especially considering the stricter regulations like the IMO 2020 that imposed significant reduction in the sulfur content of marine fuel oils (BUNKER), requiring even more capacity to treat bottom barrel streams, especially for refiners processing heavier crude oils and putting pressure on the refining margins that are still in recovery in the post pandemic scenario.

Under this scenario, process units called bottom barrels are able to improve the quality of crude oil residue streams (vacuum residue, gas oils, etc.) or convert them to higher added value products that gain strategic importance, mainly in countries that have large heavy crude oil reserves. These process units are fundamental to complying with the environmental and quality regulations, as well as ensuring profitability and competitiveness of refiners through raising refining margins. In a most recent trend, the reduction in transportation fuel demand is leading the refiners to look for closer integration with petrochemical assets, creating a necessity to maximize the yield of petrochemicals in the refining hardware. The use of residue upgrading technologies to produce petrochemicals against transportation fuels can be an attractive route in some markets.

Available technologies to process bottom barrel streams involve processes that aim to raise the H/C ratio in the molecule, either through reducing the carbon quantity (processes based on carbon rejection) or through hydrogen addition. Technologies that involve hydrogen addition encompass hydrotreating and hydrocracking processes, while technologies based on carbon rejection refer to thermal cracking processes like visbreaking, delayed coking, and fluid coking; catalytic cracking processes like fluid catalytic cracking (FCC); and physical separation processes like solvent deasphalting units. Some refining schemes can apply the synergy between residue upgrading technologies, aiming to ensure higher added value to the processed crude oils as well as meet the regulations and ensure higher refining margins in the downstream sector. An interesting case is the combination of hydrocracking and fluid catalytic cracking technologies; normally these technologies are faced as competitors due to the feed streams that are processed, but in some cases, the synergy between the hydrocracking and FCC technologies can ensure high bottom barrel conversion capacity and profitability, despite the relatively high capital investment.

The continuous supply of adequate crude oil to the refining hardware is one of the assumptions adopted by the refiners for the installation of refining assets or economic

DOI: 10.1201/9781003472216-13

analysis of already installed units. However, according to the installed geopolitical scenario, the supply of adequate crude oil to the refining hardware can be seriously threatened, mainly by refiners that operate with lighter and high cost crudes.

Considering the arguments presented above, more flexible refining hardware in relation to the processed crude slate is an important competitive advantage in the downstream sector, mainly the processing of heavy and extra-heavy crudes due to their lower acquisition cost when compared with the lighter crude oils. The difference in the acquisition cost between these oils is based on the yield of high added value streams that these oils present in the distillation process. Since the lighter crudes normally show higher yields of distillates than the heavier crudes, their market value tends to be higher.

Normally, the valuation of crudes is defined by the quality and the available market. In other words, it's necessary to find refiners capable of processing this crude oil and the capacity to transport the crude oil to the consumer market. Heavier crudes tend to present discounts related to lighter crude due to these three variables:

1. Quality: Heavier crudes present a lower yield of distillates and high added value derivatives like diesel, kerosene, and gasoline than lighter crudes.
2. Consumer market: The refiners able to process heavier crudes need to rely on adequate bottom barrel conversion capacity; in other words, more complex refineries, restricting the consumer market in comparison with lighter crudes.
3. Transportation: Heavier crudes present higher logistics costs due to higher energy consumption.

Despite these characteristics, refiners with adequate refining hardware and easy access to heavier crudes can use the price gap between light and heavy crude as an opportunity to improve the refining margins, mainly considering stricter regulations like the IMO 2020 that reduced, even more, the acquisition cost of heavier and sourer crudes.

13.1.1 Fluid Catalytic Cracking Technologies – The Carbon Rejection Route

The installation of catalytic cracking units allows the refiners to process heavier crude oils and consequently cheaper, raising the refining margin, mainly in higher crude oil price scenarios or in geopolitics crises that can make access to light oils difficult. The typical catalytic cracking unit feed stream is gas oils from the vacuum distillation process. However, some variations are found in some refineries, like sending heavy coke naphtha, coke gas oils, and deasphalted oils from deasphalting units to processing in the FCC unit.

In a conventional scheme, the catalyst regeneration process consists of the carbon partial burning deposited over the catalyst, according to the chemical reaction below:

$$C + \frac{1}{2} O_2 \rightarrow CO$$

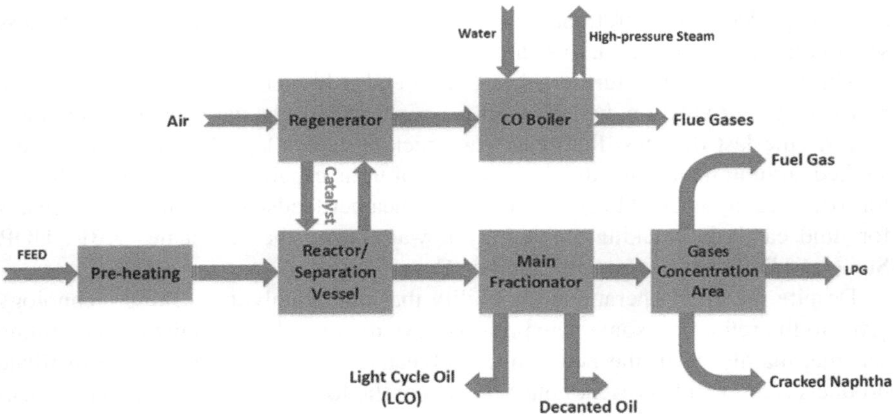

FIGURE 13.1 Schematic process flow for a typical fluid catalytic cracking process unit (FCCU).

The carbon monoxide is burned in a boiler capable of generating higher pressure steam that supplies other process units in the refinery. Figure 13.1 presents a process scheme for a typical fluid catalytic cracking unit (FCCU).

The principal operational variables in a fluid catalytic cracking unit are reaction temperature, normally considered the temperature at the top of the reactor (called a riser), feed stream temperature, feed stream quality (mainly carbon residue), feed stream flow rate, and catalyst quality. Feedstock quality is especially relevant, but this variable is a function of the crude oil processed by the refinery, so it is difficult to change; for example, aromatic feedstock with high metals content is refractory to cracking and conducting to quick catalyst deactivation.

An important variation of the fluid catalytic cracking technology is the residue fluid catalytic cracking unit (RFCC). In this case, the feedstock to the process is basically residue from the atmospheric distillation column, due to the high carbon residue and contaminants (metals, sulfur, nitrogen, etc.) are necessary for some adaptations in the unit, like a catalyst with higher resistance to metals and nitrogen and catalyst coolers. Furthermore, it's necessary to apply materials with the most noble metallurgy due to the higher temperatures reached in the catalyst regeneration step (due to the higher coke quantity deposited on the catalyst), which raises significantly the capital investment to the unit installation. Nitrogen is a strong contaminant to the FCC catalyst because it neutralizes the acid sites of the catalyst, which are responsible for the cracking reactions.

When the residue has high contaminant content, it is common for the feed stream treatment in hydrotreating units to reduce the metals and heteroatom concentration to protect the FCC catalyst. Typically, the average yield in fluid catalytic cracking units is 55% in volume in cracked naphtha and 30% in LPG.

Usually, catalytic cracking units are optimized to aim the production of fuels (mainly gasoline); however, some process units are optimized to maximize the light olefins production (propylene and ethylene). Process units dedicated to this purpose have his project and operational conditions significantly changed once the process severity is strongly raised in this case.

The reaction temperature reaches 600°C and a higher catalyst circulation rate raises the gas production, which requires a scaling up of the gas separation section.

Over the last decades, fluid catalytic cracking technology has been intensively studied, aiming mainly for the development of units capable of producing light olefins (deep catalytic cracking) and processing heavier feedstocks. The main licensers for fluid catalytic cracking technology nowadays are the companies KBR, UOP, Stone and Webster, Axens, and Lummus Company.

Despite the great operational flexibility that fluid catalytic cracking technology gives to the refineries, some new projects have dismissed these units in the refining scheme, mainly when the new refinery objective is to maximize middle distillate products (diesel and kerosene) once this is not the focus of the fluid catalytic cracking unit.

13.1.2 Improving the Yield of Petrochemicals in the Refining Hardware – The Petrochemical FCC Technologies

As quoted earlier, in markets with high demand for petrochemicals, the petrochemical FCC can be an attractive alternative to refiners, aiming to ensure higher added value to bottom barrel streams. An example of FCC technology developed to maximize the production of petrochemical intermediates is the PetroFCC™ process by UOP Company. This process combines a petrochemical FCC and separation processes optimized to produce raw materials for the petrochemical process plants. Other available technologies are the HS-FCC™ process commercialized by Axens Company and the INDMAX™ process licensed by Lummus Company.

For petrochemical FCC units, the reaction temperature reaches 600°C and a higher catalyst circulation rate raises the gas production, which requires a scaling up of the gas separation section. The higher thermal demand makes operating the catalyst regenerator advantageous, leading to the necessity of installing a catalyst cooler system.

The installation of petrochemical catalytic cracking units requires a deep economic study taking into account the high capital investment and higher operational costs; however, some forecasts indicate growth of 4.0% per year to the market of petrochemical intermediates until 2025. In this scenario, the capital investment aiming to raise the market share in the petrochemical sector can be attractive, allowing then a favorable competitive positioning for the refiner through the maximization of petrochemical intermediates.

In refining hardware with conventional FCC units, further than the higher temperature and catalyst circulation rates, it's possible to apply the addition of catalyst additives like the zeolitic material ZSM-5 that can raise the olefins yield close to 9.0% in some cases when compared with the original catalyst. This alternative raises the operational costs; however, as aforementioned, it can be economically attractive considering the petrochemical market forecasts. Figure 13.2 presents some optimization strategies to improve the petrochemical yield in conventional FCC units.

The use of FCC catalyst additives such as ZSM-5 can increase unit propylene production by up to 8.0%.

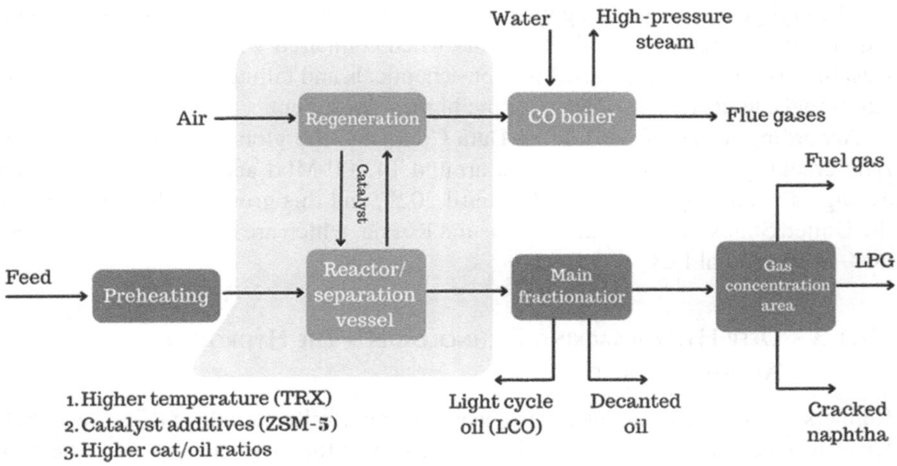

FIGURE 13.2 Optimization of process variables in FCC units to improve the yield of petrochemical intermediates.

The installation of a catalyst cooler system raises the process unit's profitability through the total conversion enhancement and selectivity to the noblest products, such as propylene and naphtha, against gases and coke production. The catalyst cooler is necessary when the unit is designed to operate under total combustion mode due to the higher heat release rate as presented below:

$$C + ½O_2 \rightarrow CO\left(\text{Partial Combustion}\right)\ \Delta H = -27\,\text{kcal/mol}$$

$$C + O_2 \rightarrow CO_2\left(\text{Total Combustion}\right)\ \Delta H = -94\,\text{kcall/mol}$$

In this case, the temperature of the regeneration vessel can reach values close to 760°C, leading to higher risks of catalyst damage, which is minimized through catalyst cooler installation. The option by the total combustion mode needs to consider the refinery thermal balance, as, in this case, it will not have the possibility to produce steam in the CO boiler. Furthermore, the higher temperature in the regenerator requires materials with the noblest metallurgy. This significantly raises the installation costs of these units, which can be prohibitive to some refiners with restricted capital access.

Among other petrochemical FCC technologies, it's possible to quote the Maxofin™ and K-COT™ processes developed by KBR Company and the SCC™ technology developed by Lummus Company.

Due to the higher production of light olefins, mainly ethylene, another important difference between conventional and petrochemical FCC units is related to the gas recovery section. While conventional FCC applied absorber columns, in petrochemical units, cryogenic processes are applied through refrigeration cycles in similar conditions that are applied in steam cracking units.

The cryogenic processes applied to olefin recovery raise, even more, the capital requirement to petrochemical FCC units when compared with conventional FCCs. Despite this, the growing market for petrochemicals and falling demand for transportation fuels, tends to compensate for the higher investment.

According to data from Global Data Company, the global installed fluid catalytic cracking capacity in 2022 was around 14,000 Mbd and will grow under an average annual growth rate of 3.0% until 2027, and this growth will be headed by the United States, China, India, Japan, and Russia, which are responsible for around 66% of the global FCC.

13.1.3 Deep Hydrocracking Technologies – The Hydrogen Addition Route

Refiners processing heavy and extra-heavy (or high sulfur) crudes face a great challenge to meet the IMO 2020, as it is extremely difficult to comply with the new regulation through carbon rejection technologies. In this case, the hydrogen addition technologies are fundamental.

The hydroprocessing of residual streams presents additional challenges when compared with the treating of lighter streams, mainly due to the higher contaminants content and residual carbon (RCR) related to the high concentration of resins and asphaltenes in the bottom barrel streams.

Higher metals and asphaltenes content led to a quick deactivation of the catalysts through high coke deposition rate, catalytic matrix degradation by metals like nickel and vanadium, or even by the plugging of catalyst pores produced by the adsorption of metals and high molecular weight molecules on the catalyst surface. For this reason, according to the content of asphaltenes and metals in the feed stream, more versatile technologies are adopted, aiming to ensure an adequate operational campaign and an effective treatment.

Despite the high performance, the fixed bed hydrocracking technologies can be not economically effective in treating residue from heavy and extra-heavy due to the short operating lifecycle. Technologies that use ebullated bed reactors and continuum catalyst replacement allow higher campaign periods and higher conversion rates. Among these technologies, the most known are the H-Oil and Hyvahl™ technologies developed by Axens Company, the LC-Fining process by Chevron-Lummus, and the Hycon™ process by Shell Global Solutions. These reactors operate at temperatures above 450°C and pressures of 250 bar.

Catalysts applied in hydrocracking processes can be amorphous (alumina and silica-alumina) and crystalline (zeolites) and have bifunctional characteristics, once the cracking reactions (in the acid sites) and hydrogenation (in the metals sites) occur simultaneously.

An improvement in relation to ebullated bed technologies is the slurry phase reactors, which can achieve conversions higher than 95%. In this case, the main available technologies are the HDH™ process (Hydrocracking-Distillation-Hydrotreatment), developed by PDVSA-Intevep; the VEBA-Combicracking process (VCC)™ commercialized by KBR Company; the EST™ process (Eni Slurry Technology) developed by the Italian state oil company ENI; and the Uniflex™ technology developed by UOP Company.

In the slurry phase hydrocracking units, the catalysts are injected with the feedstock and activated in situ while the reactions are carried out in slurry phase reactors, minimizing the reactivation issue and ensuring higher conversions and operating lifecycles.

Other commercial technologies for slurry hydrocracking processes are the LC-Slurry™ technology developed by Chevron Lummus Company and the Microcat-RC™ process by Exxon Mobil Company. Aiming to meet the new bunker quality requirements, the noblest streams, normally directed to produce middle distillates, can be applied to produce low sulfur fuel oil. This can lead to a shortage of intermediate streams to produce these derivatives, raising their prices. The market for high sulfur content fuel oil should be strongly reduced, due to the higher price gap when compared with diesel; its production tends to be economically unattractive.

According to data from Global Data Company, the global installed hydrocracking capacity in 2022 was around 12,500 Mbd and will grow under an average annual growth rate of 5.0% until 2027, and this growth will be headed by the United States, China, India, and Saudi Arabia.

13.1.4 Petrochemical Integration – Strong Dependence of High Bottom Barrel Conversion

The focus of the closer integration between refining and petrochemical industries is to promote and seize the synergies existing opportunities between both downstream sectors to generate value for the whole crude oil production chain. Table 13.1 presents the main characteristics of the refining and petrochemical industry and the synergies potential.

As aforementioned, the petrochemical industry has been growing at considerably higher rates when compared with the transportation fuels market in the last years; additionally, it represents the noblest destiny and is less environmentally aggressive to crude oil derivatives. The technological bases of the refining and petrochemical industries are similar, which leads to possibilities of synergies, capable of reducing operational costs and adding value to derivatives produced in the refineries.

TABLE 13.1
Refining and Petrochemical Industry Characteristics

Refining Industry	Petrochemical Industry
Large feedstock flexibility	Raw material from naphtha/NGL
High capacities	Higher operation margins
Self-sufficient in power/steam	High electricity consumption
High hydrogen consumption	High availability of hydrogen
Streams with low added value (unsaturated gases and C2)	Streams with low added value (heavy aromatics, pyrolysis gasoline, C4's)
Strict regulations (benzene in gasoline, etc.)	Strict specifications (hard separation processes)
Transportation fuels demand in declining at global level	High-demand products

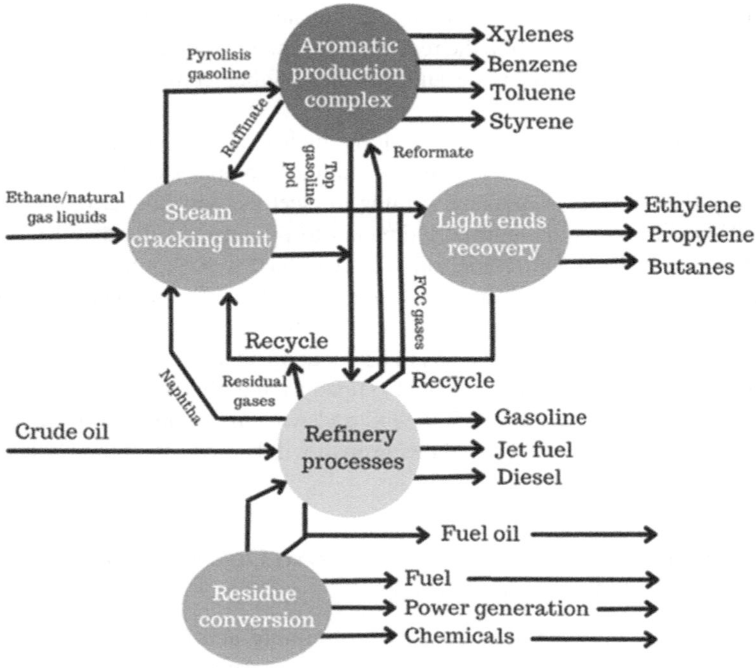

FIGURE 13.3 Synergies between refining and petrochemical processes.

Figure 13.3 presents a block diagram that shows some integration possibilities between refining processes and the petrochemical industry.

Process streams considered with low added value to refiners, like fuel gas (C2), are attractive raw materials to the petrochemical industry, as well as streams considered residual to petrochemical industries (butanes, pyrolysis gasoline, and heavy aromatics), which can be applied to refiners to produce high-quality transportation fuels. This can help the refining industry meet the environmental and quality regulations for derivatives.

The integration potential and the synergy among the processes rely on the refining scheme adopted by the refinery and the consumer market. Process units such as fluid catalytic cracking (FCC) and catalytic reforming can be optimized to produce petrochemical intermediates to the detriment of streams that will be incorporated into the fuel pool. In the case of FCC, the installation of units dedicated to producing petrochemical intermediates, called petrochemical FCC, aims to reduce to a minimum the generation of streams to produce transportation fuels; however, the capital investment is high once the severity of the process requires the use of material with the noblest metallurgical characteristics.

The IHS Markit Company proposed a classification of the petrochemical integration grades, as presented in Figure 13.4.

According to the classification proposed, the crude-to-chemicals refineries are considered the maximum level of petrochemical integration.

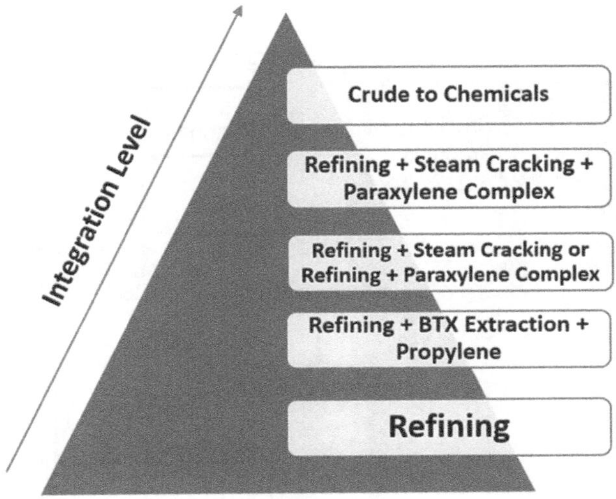

FIGURE 13.4 Petrochemical integration levels (Based on IHS Markit, 2018).

13.1.5 The Crude Oil to Chemicals – What Is the Role of FCC and Hydrocracking?

Due to the increasing market and higher added value, as well as the trend of reduction in transportation fuel demand, some refiners and technology developers have dedicated their efforts to developing crude-to-chemicals refining assets. One of the big players that have been invested in this alternative is the Saudi Aramco Company. The concept is based on the direct conversion of crude oil to petrochemical intermediates, as presented in Figure 13.5.

The process presented in Figure 13.5 is based on the quality of the crude oil and deep conversion technologies like high severity or petrochemical FCC units and deep hydrocracking technologies. The processed crude oil is light with low residual carbon, which is a common characteristic in Middle Eastern crude oils. The processing scheme involves a deep catalytic conversion process, aiming to reach maximum conversion to light olefins. In this refining configuration, the combination of hydrocracking and petrochemical FCC is applied to ensure maximum added value to the processed crude oil through near zero production of transportation fuels.

13.1.6 Synergy between Hydrocracking and FCC Units – A High Bottom Barrel Conversion Refining Hardware

As aforementioned, sometimes the hydrocracking and FCC technologies are faced with competitors' technologies in the refining hardware due to the similarities of feed streams that are processed in these units. In some refining schemes, the mild hydrocracking units can be applied as pretreatment steps to FCC units, especially to bottom barrel streams with high metal content that are severe poisons to FCC catalysts. Furthermore, the mild hydrocracking process can reduce the residual carbon to

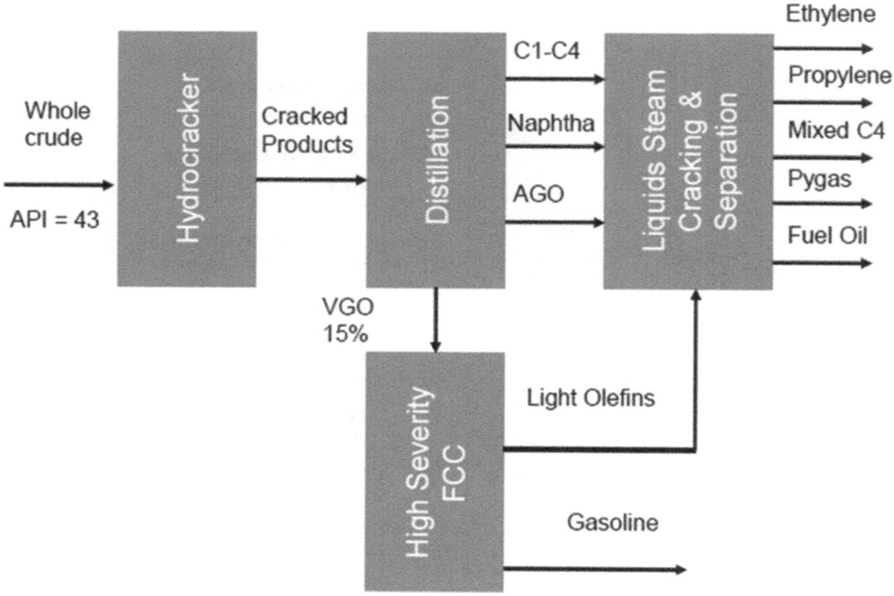

FIGURE 13.5 Saudi Aramco crude oil to chemicals concept (IHS Markit, 2017).

FCC feed, raising the performance of the FCC unit and improving the yield of light products like naphtha, LPG, and olefins.

The integrated refining scheme relies on deep residue upgrading technologies, as hydrocracking and fluid catalytic cracking (FCC) are capable of reaching the production of high-quality petrochemicals, according to the market trends. In some refining configurations, mild hydrocracking technologies are applied upstream of FCC units aiming to improve the quality of the FCC products, mainly the sulfur content.

Considering the great flexibility of deep hydrocracking technologies that are capable of converting feed streams varying from gas oils to residue, an attractive alternative to improve the bottom barrel conversion capacity is to process the hydrocracking units with the uncracked residue in the FCC unit, aiming to improve the yield of high added value derivatives in the refining hardware, mainly middle distillates like diesel and kerosene. This configuration can be especially attractive for gasoline Tier 3 production (with a maximum of 10 ppm sulfur) for refiners processing high sulfur crudes.

As described above, the combination of FCC and hydrocracking processing units can help the refiners to maximize the added value to the processed crude through the production of high added value molecules, especially petrochemicals, considering the growing demand for petrochemical intermediates like ethylene and propylene in detriment to gasoline, which present demand in contraction. Figure 13.6 presents another refining configuration based on crude-to-chemicals concept; again, the refining scheme relies on FCC and hydrocracking synergy.

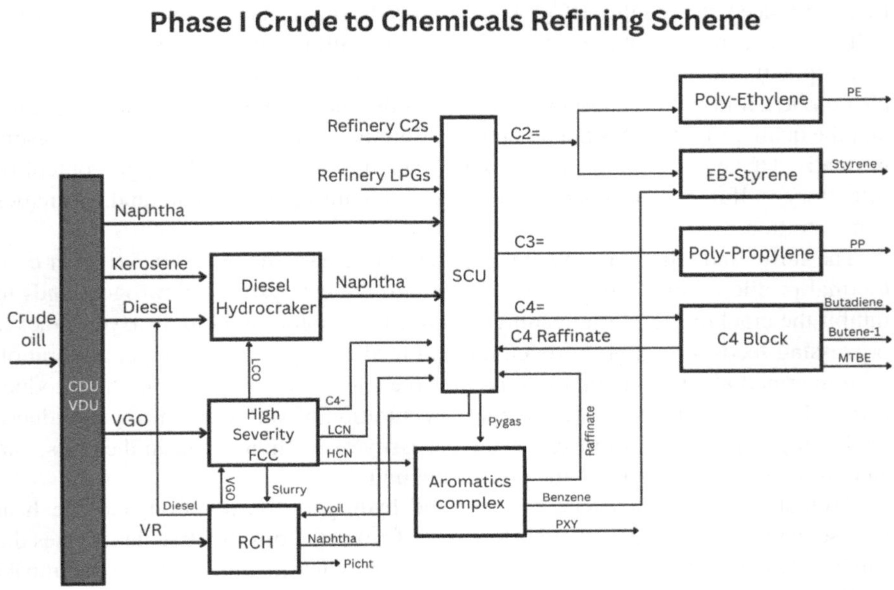

FIGURE 13.6 Typical refining configuration of crude to chemicals refinery.

13.1.7 The Side Effect of Cracked Feeds – A Special Challenge to Hydrocracking Units

The most common cracked feeds directed to hydrocracking units are residual streams from FCC, like light cycle (LCO) and decanted oil (DO), and heavy coker gasoil (HCGO) from delayed coking units. Another less common feed is residue from visbreaking units.

The main characteristics that influence the hydrocracking performance for each feedstock are presented below:

- FCC cycle oils: Present high aromaticity that are normally refractory to cracking reactions as well as refractory sulfur components, raising the sulfur content in the final products and reducing diesel cetane number. On the other side, normally, it presents low basic nitrogen content that is a poison to the hydrocracking catalysts.
- Thermal cracking feeds: Normally, it presents low aromatic content but concentrates refractory sulfur components.

The HCGO is an interesting case study as a feed to a hydrocracking unit. Refiners with high complexity refining hardware can rely on the synergy between delayed coking and hydrocracking technologies to ensure added value to bottom barrel streams.

The quality of the HCGO relies on the quality of the feed to the delayed coking unit as well as the operating mode of the unit, mainly the recycling ratio. Higher recycling ratios produce better quality HCGO once reduced Conradson Carbon Residue (CCR), reducing the contaminant contents such as metals, sulfur, and nitrogen.

Despite this advantage, the delayed coking operators normally minimize the recycle ratio to a minimum as possible, aiming to raise the fresh feed processing capacity, and the quality of HCGO is not an optimization focus of the refinery. For this reason, normally, HCGO is a hard feed to hydrocracking units due to the high content of refractory sulfur components, high CCR, high nitrogen content, and aromatics concentration.

The sulfur and nitrogen content raises the heat release in the first bed (higher exothermal profile), which can produce damage to the catalysts. The nitrogen tends to inhibit the cracking reaction, leading to lower conversion in the unit. Hydrocracker processing feeds with high nitrogen content tend to apply processing configuration with intermediate gas separation to control the catalyst activity. The higher production of H_2S and NH_3 due to the higher concentration of sulfur and nitrogen reduces the hydrogen partial pressure, raises the necessity of washing water in the units, and can raise the corrosion rate in the processing unit.

Aromatics compounds tend to raise the hydrogen consumption and the heat release in the catalyst bed and are precursors of coking deposition that deactivates the catalyst. Other side effects of the cracked feeds to hydrocracking units are the impact on the quality of the final products, like a lower cetane number of diesel, a higher smoke point of kerosene, a lower viscosity index in the lubricating oils, and higher sulfur content.

As described above, processing cracked feeds in hydrocracking units presents some additional challenges to refiners related to hydrogen consumption, better quench design of the catalyst bed due to the higher exothermic profile of the reactions, and lower global activity of the catalyst due to the higher poison content, like basic nitrogen. These characteristics lead the refiners processing cracked feeds in hydrocracking units to invest more capital in feed-treating systems like filtering and guarding beds. Despite this apparent disadvantage, refiners able to add value to bottom barrel streams can enjoy a highly competitive advantage considering the downstream market post-IMO 2020. For refiners processing extra-heavy bottom barrel streams, deep hydrocracking technologies like slurry phase hydrocracking can be an interesting option, despite the high capital and operating costs.

It is expected that there will be significant growth in the global hydrocracking installed capacity in the next years. According to data from Global Data Company, the global installed hydrocracking capacity in 2022 was around 13,000 Mbd and will grow under an average annual growth rate of 5.0% until 2027, and this growth will be headed by the United States, China, India, Japan, Russia, and Saudi Arabia, which are responsible for around 56% of the global hydrocracking capacity.

The headed by Asian players is expected once these players present high integration levels between refining and petrochemical assets, requiring high bottom barrel conversion capacity to maximize the yield of petrochemicals. Again, this shows the competitive advantage for the Asian players due to the highest flexibility and profitability of their refining hardware.

As described above, the fluid catalytic cracking units can ensure operational flexibility and higher refining margins for the downstream players under the challenge scenario imposed on the fossil fuels and are expected to increase investments in FCC installations for the next years. Based on data from Global Data Company, the global installed fluid catalytic cracking capacity will rise from 14.4 million barrels per day (MMbpd) in 2022 to 15.8 MMbpd in 2026, a growth of 9.3% in four years.

As expected, the growth of FCC capacity will be led by the Asian players, who present the highest integration capacity among the downstream players and have been investing heavily to maximize the petrochemical yield in their refining assets, including crude-to-chemicals refineries. The Asian players will add close to 900 thousand barrels per day (Mbd) to the global FCC processing capacity until 2026, followed by the African players, who have invested in FCC as part of their strategy to reduce the external dependence on high-quality gasoline. In other words, unlike Asian players, the addition of global FCC capacity produced by African players tends to be related to conventional and low severity FCC units, with the gasoline production focus especially led by the new Dangote refinery in Nigeria. The African players will add close to 570 Mbd to the global FCC processing capacity until 2026.

The current scenario indicates a competitive advantage to refiners capable of maximizing the yield of petrochemicals in the refining hardware, ensuring higher added value to the processed crude oils, and the adequate synergy between refining technologies is a basic concept in the downstream sector and one of the first steps to define adequate refining configuration. The synergy between residue upgrading technologies is increasingly relevant to the refiners, aiming to keep and improve the economic sustainability of the refiners, especially considering the downstream market post IMO 2020. Considering the limitations of FCC technology to treat heavier feedstocks, the synergy with hydrocracking and mild hydrocracking units can allow a higher level of bottom barrel conversion and maximum production of high added value derivatives.

Despite the relatively high capital cost, the synergy between hydrocracking and FCC units is economically attractive to the refiners inserted in markets with higher demand for lighter derivatives, especially petrochemicals and middle distillates. Despite the benefits of petrochemical integration, it's fundamental to take into account the necessity to reach a circular economy in the downstream industry. To achieve this goal, the chemical recycling of plastics is essential. As presented above, there are promising technologies which can ensure the closing of the sustainability cycle of the petrochemical industry.

13.2 MEETING THE GASOLINE QUALITY AND PROCESS SAFETY REQUIREMENTS THROUGH NEW NAPHTHA ALKYLATION TECHNOLOGIES

The process accident that occurred in PES (Philadelphia Energy Solutions) Company in June of 2019 restarts the discussion about the security of naphtha alkylation units based on HF (hydrofluoric acid) as a homogeneous catalyst. In the last years, some refiners have decided to stop the operation of their HF alkylation units due to the unacceptable safety risks.

The high toxicity and corrosion risks associated with hydrofluoric acid require great attention from the inspection equipment team to control the integrity of process equipment. When this management system fails, the consequence is drastic, like it's possible to conclude based on the previous analysis conducted by CSB (US Chemical Safety and Hazard Investigation Board).

In recent years, some process accidents involving alkylation units have called attention to the safety characteristics of these process units. Below, we quote the last process safety accidents involving alkylation units:

1. PES Refinery in USA, 2019
2. CITGO Corpus Christy Refinery in the United States, 2009
3. Hube Globe Company in South Korea, 2008
4. Marathon Petroleum Refinery the United States, 1987
5. Exxon Mobil Oil Refinery in Torrance, California, the United States, 2015
6. Ciniza Oil Refinery in the United States, 2004
7. Husky Superior Refinery in the United States, 2018

This short list of process accidents calls the attention of the authorities to the real safety risks associated with the presence of alkylation process units based on HF close to urban centers, as a great release of HF can affect the population and cause severe damages and human losses. According to CSB information, only in the United States, there are 46 HF naphtha alkylation units in operation.

It's interesting to highlight that the naphtha alkylation technology as an octane-boosting route can be attractive to refiners inserted in markets with high demand for gasoline and availability of LPG, like in the United States. An interesting case study is the Brazilian market, which historically, Brazil relied on external markets to supply their needs of LPG. In 2020, the internal production of LPG in Brazil was 9.86 million m^3 where 74% was supplied by refineries (7.34 million m^3), while petrochemical plants and natural gas processing units were responsible for the remaining internal production. In the same year, the imports of LPG in Brazil reached 3.62 million m^3. With the development of pre-salt crude oil reserves, natural gas production in Brazil is growing, and the production of LPG associated with the natural gas tends to reduce the external dependence of this derivative. According to the Brazilian Petroleum Agency (ANP), the imports of natural gas reduced 20% in 2020 in comparison with 2019, reaching 7.9 billion m^3.

This scenario can lead the Brazilian refiners to consider the naphtha alkylation route as octane boosting to produce low sulfur and high-quality gasoline at the same time, which can deal with an eventual surplus of LPG in the internal market. Considering the growing propylene gap, which is observed at the global level, another attractive route to deal with LPG surplus is the propane dehydrogenation (PDH) units that can be applied to add significant value to the LPG.

Despite the risks related to the operation of naphtha alkylation units, the alkylate is an important octane booster for gasoline, and some forecasts indicate an annual growth rate of the global alkylation units capacity of 1.5% until 2025. According to Global Data Company, the global alkylation capacity increased from 2.406 thousand barrels per day (mbpd) in 2015 to 2.553 mbpd in 2020, and the United States is responsible for 1.276 mbpd of this capacity.

As aforementioned, the United States leads the global capacity of naphtha alkylation, but China presents the higher growth of naphtha alkylation capacity between 2015 and 2020 with a growth of 28%, followed by South Korea with a growth of 3.9%. This data shows that the naphtha alkylation sector can be an attractive market for technology licensors for the next years and the relevance of naphtha alkylation processing units to compose the gasoline pool of the refineries.

13.2.1 Gasoline Production Process – The Relevance of Octane Boosters

The final gasoline is composed of a blending of different naphtha streams, as presented in Figure 13.7.

The use of straight-run and reformed naphtha is normally minimized, aiming to direct these streams to the petrochemical intermediates market due to the higher added value of these streams in this market.

Cracked naphtha (naphtha from FCC) contributes positively to the octane number of the final gasoline; however, due to the current restrictions related to the sulfur content in the gasoline (a maximum of 10 ppm), the use of cracked naphtha without treatment step is limited. Refineries that have catalytic alkylation units in their refining scheme normally direct this stream to produce aviation gasoline that has higher market value when compared with automotive gasoline. For this reason, the participation of the alkylation naphtha is minimized in the composition of this fuel.

Isomerization naphtha has low contaminant content (sulfur and nitrogen) and a high octane number, and for these reasons, the participation of this stream in the formulation of gasoline is maximized in refineries that have isomerization units in the refining scheme. In markets with high demand for gasoline, refiners can add butanes to the gasoline pool; however, the participation is limited due to the high vapor pressure of this stream that can lead to breaking quality requirements (Reid vapor pressure – RVP). Normally, butanes are added to the LPG pool, respecting the limits to avoid breaking the mixture quality related to heavy limitations.

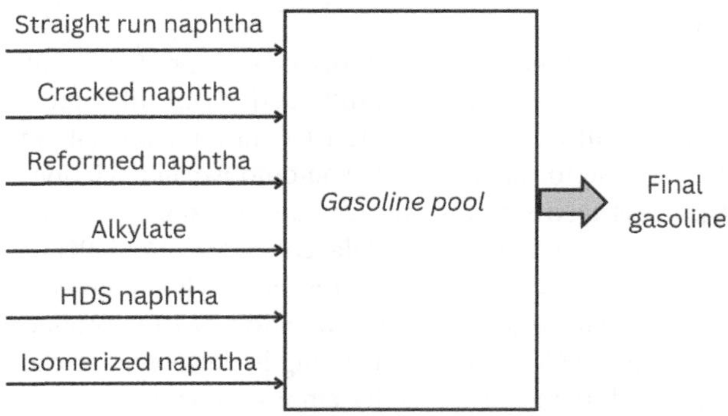

FIGURE 13.7 Example of process streams blending to produce gasoline.

Nowadays, due to the environmental and quality regulations over gasoline, the hydrodesulfurization of cracked naphtha is fundamental to be possible for the refiners to meet the sulfur content in the final gasoline that can be lower than 10 ppm in most restrictive markets. The alkylated naphtha is one of the main octane boosters in the gasoline pool and, according to the refining scheme, is a fundamental stream to refiners achieving the current quality requirements of final gasoline.

13.2.2 Traditional Technologic Routes to Naphtha Alkylation

Gasoline is one of the most consumed crude oil derivates and is normally produced through a mixture of naphtha from different refining process steps. The streams normally involved in the gasoline production process are straight-run naphtha, cracked naphtha, coke naphtha (after hydrotreatment), and reforming naphtha.

One of the main parameters of gasoline quality is the octane number, which is a measure of the combustion quality of this derivate; one of the streams that contributes to raising the octane number is the reformed naphtha, produced in the catalytic reforming unit. However, due to the severe restrictions related to the carcinogenic aromatic emissions, mainly benzene, some refiners have avoided the application of this stream to formulate gasoline, directing the reformed naphtha preferably to petrochemical intermediates production in aromatics complexes.

An alternative to the reforming naphtha is the production of branched hydrocarbons (with high octane number) through the catalytic alkylation process.

The alkylation process involves the reactions between light olefins (C_3–C_5) and isoparaffinic hydrocarbons like isobutane. The reaction product called alkylate is a mixture of branched hydrocarbons with higher molecular weight and higher octane number.

An example of a typical alkylation reaction is represented below:

$$C_4H_{10} + C_3H_6 \rightarrow C_7H_{16}\left(2,3-\text{Dimethylpentane}\right)$$

The reaction is catalyzed in strongly acidic reaction environments. The acids normally employed in industrial scale technologies are hydrofluoric acid (HF) and sulfuric acid (H_2SO_4).

The main advantage of the alkylation process is the production of a stream with a high octane number, high chemical stability, and practically free of contaminants such as nitrogen and sulfur. These characteristics turn the alkylate into a component attractive to the gasoline formulation in the automotive and aviation industries.

Alkylation process feed streams are generally obtained from LPG produced in deep conversion units, mainly FCC and delayed coking. The LPG produced in these process units has high olefin content, ideal for the alkylation process. The isobutane stream is normally obtained through the separation of LPG produced in the atmospheric distillation unit, FCC, or delayed coking in deisobutanizer towers.

As aforementioned, the acids generally employed as the homogeneous catalysts in the alkylation process are HF and H_2SO_4. Figure 13.8 presents a process flow diagram of the alkylation process catalyzed by HF.

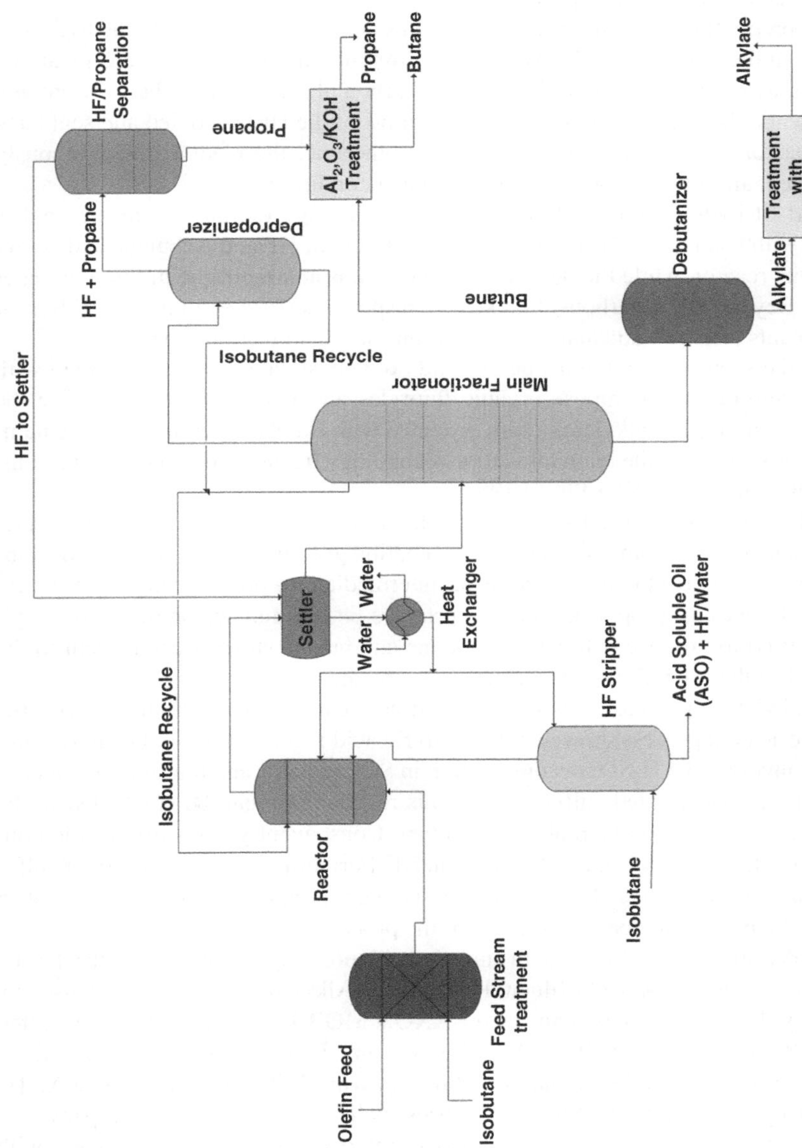

FIGURE 13.8 Typical process flow diagram to catalytic alkylation unit using HF as catalyst.

The feed stream goes through pretreatment (generally molecular sieves or alumina) before being pumped to the reactor; the objective is to remove process contaminants, mainly water, diolefins, sulfur, and nitrogen compounds. Water is especially damaging to the process, once it accelerates the piping and equipment corrosion process, and beyond that, requires higher HF reposition.

After pretreatment, the hydrocarbon streams are put in contact with the hydrofluoric acid in the reactor, and the hydrocarbon mixture and HF solution are separated through gravity in a settler vessel. The hydrocarbon phase is sent to the fractionating section while the aqueous phase (containing most of the HF) is cooled and sent back to the reactor. As alkylation reactions are exothermic, the reactor is continuously refrigerated, aiming to keep the reaction in ideal conditions.

A part of hydrofluoric acid is sent to the stripping column, where the acid is stripped with isobutane. The top product is a mixture of HF and isobutene and is sent back to the reactor, while the bottom stream contains an azeotropic mixture of water and HF. Beyond hydrocarbons, this step is responsible for keeping the HF free of contaminants and with adequate concentration for the alkylation process.

After the separation columns, butane and propane streams go to a treatment with alumina, aiming to decompose organic fluorides and with KOH to neutralize the remaining acidity. The alkylate stream is treated with NaOH to neutralize the remaining acidity. Currently, the main alkylation technology licensors with HF are the companies UOP and CONOCO-PHILLIPS.

Alkylate stream is normally directed to the refinery gasoline pool for the production of high octane automobile gasoline or aviation gasoline; however, in petrochemical plants, this stream can be used as an intermediate to produce ethylbenzene (to produce styrene), isopropyl-benzene (to produce phenol and acetone), and dodecylbenzene used to produce detergents. Propane and butane streams can be sent to the LPG pool of the refinery or commercialized separately.

The alkylation process with sulfuric acid as a catalyst has similarities with the hydrofluoric acid process; however, the sulfuric acid regeneration step is more complex and involves the H_2SO_4 decomposition in SO_2 and SO_3 and the subsequent condensation of concentrated sulfuric acid. This regeneration can be conducted in the processing site or in an external process plant. Consequently, the sulfuric acid consumption in the process is much higher than HF. Furthermore, the solubility of sulfuric acid in hydrocarbons is lower, requiring greater agitation to maintain the contact between the phases in ideal conditions for the process.

The alkylation technologies with sulfuric acid most applied on an industrial scale are the processes STRATCO Effluent Refrigerated Alkylation Process™, licensed by STRATCO Engineering Company, and EXXON MOBIL Cascade Auto refrigerated Process™, licensed by EXXON MOBIL Company. Figure 13.9 shows a simplified process flow diagram of the alkylation technology with H_2SO_4, licensed by STRACTO Engineering. Nowadays, the Stratco™ process is licensed by Dupont Company.

Olefins feed stream goes to a coalescer to remove water; after the mixture with isobutane is recycled, in the sequence, the mixture is sent to the reactor. The mixture of hydrocarbons and acid follows a settler where the phase separations occur; the organic phase is sent back to the reactor, and a control valve promotes the necessary pressure reduction to vaporize the lighter hydrocarbons and remove heat from the

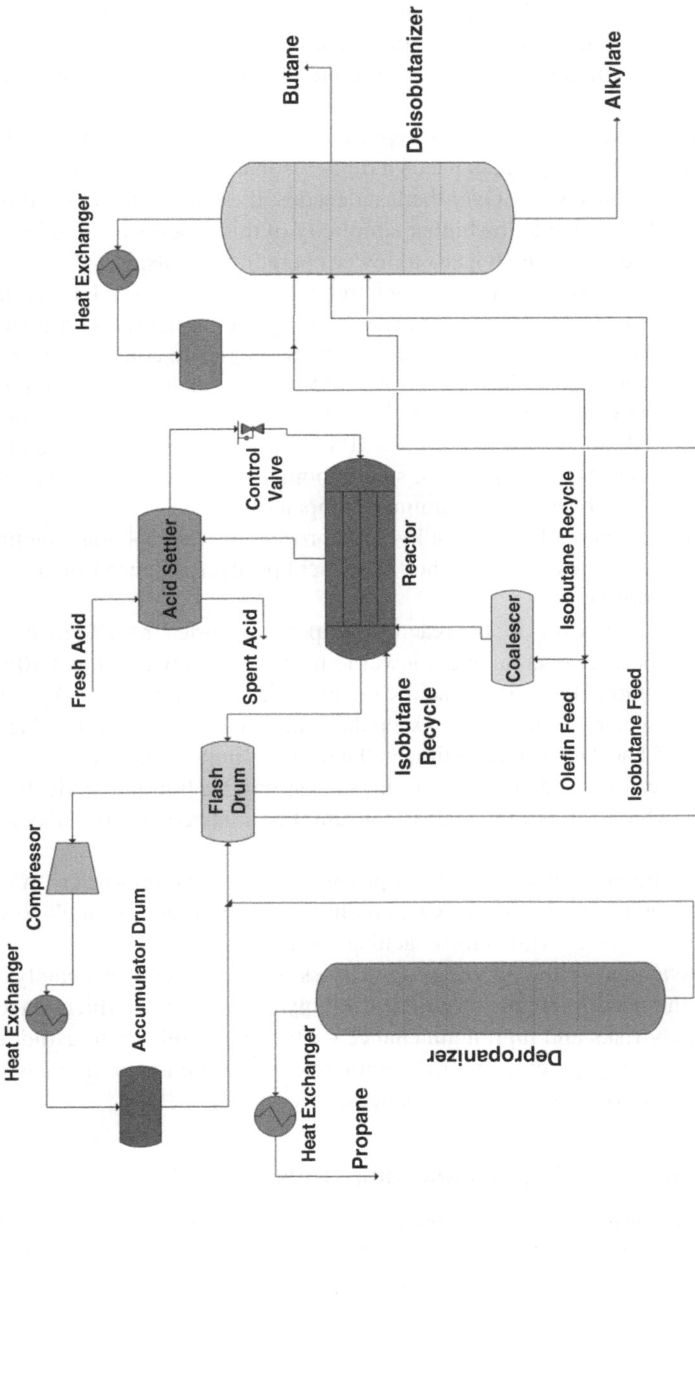

FIGURE 13.9 Basic process flow diagram of the alkylation process catalyzed by H_2SO_4, developed by STRATCO Engineering Company.

reactor, controlling the equipment temperature, which rises due to the exothermic characteristics of alkylation reactions.

The hydrocarbons blend is sent to a flash drum where the lighter phase is directed to a compressor to condense in an accumulator vessel, and the propane is recovered in the depropanizer tower while the heavier hydrocarbons (essentially isobutane) are recycled to the reactor. The stream containing the alkylate is directed to a caustic treatment and posteriorly to a deisobutanizer column where the alkylate is removed at the bottom.

As aforementioned, the need for catalyst replacement is higher in the process with sulfuric acid; however, the HF process needs a higher isobutane/olefins ratio, which means a greater separation system. Over the last decades, the refiners have opted for HF alkylation technologies due to the higher simplicity of this process and the lower need for catalyst replacement, which leads to lower operational costs.

However, regulatory pressures have led some refiners to convert their HF alkylation units to operate with H_2SO_4. Due to the high volatility and higher risks presented by the hydrofluoric acid, some licensers developed technologies to convert HF units to operate with sulfuric acids, like the ALKYSAFE™ technology, licensed by STRATCO Engineering Company, and the ReVap™ process, developed by the companies EXXON MOBIL and CONOCO-PHILLIPS, which uses additives to reduce the HF volatility, making the unit operation safer; another technology which can be quoted is the CDAlky™ process from Lummus Company.

The principal process variables of the alkylation process are the isobutane/olefins ratio, reaction temperature, acid/hydrocarbon ratio, acid purity, residence time in the settler, and operation pressure.

For alkylation processes with HF, the reaction temperature varies from 20 to 40°C, while sulfuric acid processes operate under lower temperatures, between 4 and 10°C. Under higher temperatures, the sulfuric acid can suffer decomposition to SO_2–SO_3. Operating pressure is generally sufficient to keep the hydrocarbons in the liquid phase, normally lower than 5 bar. The residence time in the settler is important because if this time is quite low, undesirable reactions can occur, such as organic fluoride production. In the case of HF, which raises the catalyst consumption and reduces the alkylate production.

Acid purity must be maintained as high as possible through the removal of ASO (acid soluble oil), water, and dissolved reactants in the HF case and through fresh acid replacement in processes with sulfuric acid as the catalyst.

The main disadvantage of the alkylation processes with homogeneous catalysts (HF or H_2SO_4) is the need to handle strong acid highly concentrated, which leaves greater process safety risks and high maintenance costs, mainly related to avoiding corrosion in piping and equipment, and, as aforementioned, equipment failures with contention losses can have dramatic consequences.

13.2.3 Alternatives and Safer Technologies to Naphtha Alkylation

Aiming to eliminate these risks, some licensors have dedicated their efforts to developing heterogeneous catalysts that can replace the strong acids in the alkylate production processes. The UOP Company developed a process called Alkylene™, which applies a solid catalyst with continuous regeneration during the process. Among other technologies, the processes LURGI EUROFUEL™, developed by Lurgi Engineering Company in cooperation with SUD-CHEMIE, the ALKYCLEAN™ process, developed by the companies ABB-Lummus and Akzo Nobel, and the process FBA™, developed by Haldor Topsoe, can be cited.

Another attractive technology to alkylation developed by UOP Company is the ISOALKY™ based on the ionic liquid catalyst, and it's possible to revamp the conventional liquid acid units to apply the new technologies. The heterogeneous catalyst applied in the substitution of liquid acids is based on zeolites and, as presented above, can offer higher operational safety to refiners and adjacent communities once it eliminates not only the process risk but also all logistics and handling of strong acids like HF and H_2SO_4. Even refiners with low capacity of capital investment are looking to change the HF by sulfuric acid (H_2SO_4) as the catalyst in their processes once the sulfuric acid is present with lower toxicity than hydrofluoric acid (HF). Another interesting technology dedicated to producing alkylated naphtha through a safer route is the Alkysafe™ process, also developed by UOP Company, which applies ionic liquids as homogenous catalysts against strong acids in conventional processes.

The Lummus Company also developed with partners a solid catalyst naphtha alkylation process called AlkyClean™.

Despite the effectiveness of the naphtha alkylation processes based on strong acids, the large number of incidents and accidents involving these units is leaving governments and regulators to raise the pressure on the refiners relying on these units, despite the significant capital spending. As quoted above, there are technical alternatives to produce alkylated naphtha in a safer manner.

The revamp of brown route naphtha alkylation processing units can be an attractive area for capital investments in the refining industry in the next few years, aiming to change the conventional homogeneous catalysts HF and H_2SO_4 for solid catalysts. Considering only in the United States, there are around 46 naphtha alkylation units, which can offer an attractive market for technology developers and licensors.

As expected, alkylates have great participation in the gasoline pool in the North American market. This is especially relevant considering the growing pressure to reach low sulfur and high octane gasoline, raising even more the relevance of alkylation processing units to comply with the necessity of high-quality and cleaner gasoline.

The catalytic alkylation process is very attractive to countries with high gasoline consumption, and there is a great availability of LPG, as in the United States. The production of specific fuels like aviation gasoline still depends on the alkylation catalytic units, so the new and safer technologies can be attractive to refiners with significant demand for this derivative.

As previously mentioned, the alkylation process offers an alternative to produce gasoline with a high octane number, with low contaminant content and without the emission of harmful compounds to human health, as in the case of naphtha from the catalytic reforming process.

Despite the high capital investment and operational costs associated with the new technologies of catalytic alkylation units, these processes can be very attractive to the refiners which operate in markets where the demand for high-quality gasoline and petrochemical intermediates is high, and it's always understood that the lower costs of HF units do not compensate for the risks of a severe process accident as occurred in PES refineries. In this case, the refining complex was shutting down, and the PES company ended its activities, and human life was exposed to risks.

Despite the risks associated with the handling of strong acids like HF and H_2SO_4, the naphtha alkylation units are still fundamental in some markets to sustain the demand for high-quality gasoline and are expected to grow the global installed capacity of naphtha alkylation. As described in this article, there are available technologies capable of ensuring an adequate balance between process safety and gasoline quality, and this market can be attractive for technology developers in the short term.

BIBLIOGRAPHY

Advances in Catalysis for Plastic Conversion to Hydrocarbons, The Catalyst Group (TCGR), 2021.

Chang, R.J. Crude Oil to Chemicals – *Industry Developments and Strategic Implications*, Presented at *Global Refining & Petrochemicals Congress* (Houston, USA), 2018.

Encyclopedia of Hydrocarbons (ENI), Volume II, Refining and Petrochemicals (2006).

Factual Update, *Fire and Explosions at Philadelphia Energy Solutions Refinery Hydrofluoric Acid Alkylation Unit*, U.S. Chemical Safety and Hazard Investigation Board (CSB), 2019.

Gary, J. H.; Handwerk, G. E. *Petroleum Refining – Technology and Economics*. 4th ed. Marcel Dekker, 2001.

Maller, A.; Gbordzoe, E. *High Severity Fluidized Catalytic Cracking (HS-FCC™): From concept to commercialization*, Technip Stone & Webster Technical Presentation to REFCOMM™, 2016.

Myers, R.A. *Handbook of Petroleum Refining Processes*. 3rd ed. McGraw-Hill, 2004.

Robinson, P.R.; Hsu, C.S. *Handbook of Petroleum Technology*. 1st ed. Springer, 2017.

Sarin, A.K. *Integrating Refinery with Petrochemicals: Advanced Technological Solutions for Synergy and Improved Profitability*, Presented at *Global Refining & Petrochemicals Congress* (Mumbai, India), 2017.

Silva, M. W. *More Petrochemicals with Less Capital Spending*. PTQ Magazine, 2020.

Singh, V.P. *KBR Olefins Technology Solutions – The Key is Flexibility*. Technical Presentation KBR Company, 2018.

Speight, J.G. *Heavy and Extra-Heavy Oil Upgrading Technologies*. 1st ed. Elsevier Press, 2013.

US Leads the Global Refinery Alkylation Units Capacity. Global Data Energy, 2021.

Vu, T.; Ritchie, J. *Naphtha Complex Optimization for Petrochemical Production*, UOP Company, 2019.

Zhang, Z. *Crude oil to chemicals: Challenges and Opportunities in a Sustainable World*, Wood Company Presentation at METECH 2024, 2024.

Zhu, F.; Hoehn, R.; Thakkar, V.; Yuh, E. *Hydroprocessing for Clean Energy – Design, Operation, and Optimization*. 1st ed. Wiley Press, 2017.

Index

Pages in *italics* refer to figures.

A

alkylation, 36, 97, 120
ammonia, 61, 72, 124
API grade, 19, 118
aromatics, 36, 47, 53
asphalt, 19, 21
asphaltene, 228
asset management, 87, 144–147, 161
atmospheric distillation, 16
atmospheric residue, 17, 119
atmospheric tanks, 156

B

benzene, 229, 238–240
biofuels, 7–11
blanketing system, 92
blending, 105–119
boiling point, 18
bottom barrel, 128
bund, 91, *95*
bunker, 55, 112–118
business strategy, 3–6

C

carbon residue, 205
catalyst, 30–41
catalyst deactivation, 57, 215
catalytic conversion, 28–30
catalytic cracking, 30
cavitation, 74
centrifugal pump, 73–75
chemical flare, 139
Claus process, 134–136
coke deposition, 119–120
coker naphtha, 23
corrosion, 128, 139, 164–170
corrosion under insulation, 174
crude oil, 16–53
crude oil derivatives, 52
crude to chemicals, 233

D

dead-legs, 178–180
deasphalting, 101
decarbonization, 7–11
delayed coking, 68
desulfurization, 54
dewatering, 89
dewaxing, 12
diesel, 10, 54–58
diesel hydrotreating, 53
distillation, 16–20
draining, 89

E

economic sustainability, 235
emissions, 45, 56
energy management, 194
environmental impact, 7–17, 94
ETBE, 45
etherification, 44
ethylene, 85, 197
extra-heavy crude, 224

F

FCC units, 25–30
fixed roofs, 81–84
flare, 121–130
flaring systems, 121
floating roofs, 80–82
fluid catalytic cracking, 196
fluid hammer, 187
fractionation tower, 18, 192
fuel gas, 206
fuel oil, 181, 194

G

gasoline, 117, 182
gauging system, 89, 202
global energetic matrix, 13–14
gravity, 92

H

heavy crude oil, 55
HF acid, 235–240
HVO, 182
hydrocracking, 56–58
hydrodesulfurization, 46
hydrogen management, 129

hydroprocessing capacity, 109
hydrotreated naphtha, 53
hydrotreating, 11, 18, 25, 46

I

ignition system, 121
integration, 5, 154, 229
isobutene, 113
isooctane, 97
isomerization, 237

J

jet Fuel, 194
JFTOT, 98

K

kerosene, 98, 106–109

L

LCO, 206
LHSV, 54
light paraffin, 26, 44
liquid-liquid extraction, 26–28
liquid recovery, 127
low level, 115, 157
low sulfur bunker, 118
LPG, 26, 36, 80
LPG Sphere, 202
lubricating, 77, 234

M

management, 140, 196
MEA, 209
metal content, 18, 31, 231
methane, 129
methanol to olefins, 198
mild hydrocracking, 231
mixing, 85–88
MTBE, 42

N

naphtha alkylation, 235–238
naphtha isomerization, 41
naphtha hydrotreating, 50
naphthenic corrosion, 174
natural gas, 112, 138
needle coke, 25
negotiation, 165
nickel, 47
nitrogen content, 233

O

octane, 97
olefins production, 33
operating issues, 77, 91
oxygenated compounds, 45

P

paraffin dehydrogenation, 44
process safety management, 140
propane dehydrogenation, 112, 117
propylene, 213

Q

quality requirements, 94

R

refining configurations, 194, 232
refining asset, 105
refining margins, 94, 103, 194
reid vapor pressure, 42, 96, 197
renewable fuels, 7, 10
renewables coprocessing, 90
residue upgrading, 2, 60
resins, 26

S

salt content, 17
shift reaction, 12
shot coke, 24
soaker visbraking, 24
solvent deasphalting, 2, 19
sour water stripping, *48–49*, 59
sponge, coke, 24
stack sealing system, 126
steam cracking, 117
sulfur recovery, 132
sweetening, 136

T

tank farm, 69, 90
tail gas, *137*
thermal conversion, 20
toluene, 26, 33
total acid number, 175

U

Uniflex process, 58, 228
UOP, 58, 115, 226

V

vacuum distillation, 18
vanadium, 56, 62, 100
venting system, 85–86, 126
visbreaking, 20

X

xylenes, 26, 33

Z

zeolite, 30, 55, 115

For Product Safety Concerns and Information please contact our
EU representative GPSR@taylorandfrancis.com Taylor & Francis
Verlag GmbH, Kaufingerstraße 24, 80331 München, Germany